INSTRUCTOR'S GUIDE

FOR

Essentials of Human Anatomy & Physiology Laboratory Manual

THIRD EDITION

ELAINE N. MARIEB, R.N., PH.D
Holyoke Community College

This Instructor's Guide is based upon the *Human Anatomy & Physiology Laboratory Manual* (Main 7/e, Cat 8/e, Pig 8/e) Instructor's Guide, by Linda Kollett.

San Francisco Boston New York
Cape Town Hong Kong London Madrid Mexico City
Montreal Munich Paris Singapore Sydney Tokyo Toronto

Publisher: Daryl Fox
Executive Editor: Serina Beauparlant
Project Editor: Karoliina Tuovinen
Editorial Assistant: Alex Streczyn
Managing Editor: Wendy Earl
Production Editor: David Novak
Cover Design: Yvo Riezebos
Compositor: Greene Design
Senior Manufacturing Buyer: Stacey Weinberger
Executive Marketing Manager: Lauren Harp

ISBN 0-8053-7341-1
1 2 3 4 5 6 7 8 9 10—TCS—09 08 07 06 05
www.aw-bc.com

Preface

Organization of this Instructor's Guide

The Instructor's Guide for the third edition of *Essentials of Human Anatomy & Physiology Laboratory Manual* by Elaine N. Marieb features a wealth of information for the anatomy & physiology laboratory instructor.

Each exercise in this manual includes detailed directions for setting up the laboratory, comments on the exercise (including common problems encountered), some additional or alternative activities, and answers to the questions that appear in the text of the lab manual. (Answers to questions regarding student observations and data have not been included.)

Part two of the Instructor's Guide provides answers to the Review Sheets, which are offered in the laboratory manual. In some cases several acceptable answers have been provided.

The time allotment at the beginning of each exercise, indicated by the hour glass icon, is an estimate of the amount of in-lab time it will take to complete the exercise, unless noted otherwise. If you are using multimedia, add the running time to the time alloted for a given exercise.

Suggested multimedia resources, indicated by the computer icon, are listed for each exercise. Format options include VHS, CD-ROM, and DVD. The resources are also listed by system in Multimedia Resources in Appendix A (page 181) of the guide. Information includes title, distributor, running time, and format. The key to format abbreviations is on the first page of this appendix. A listing of the multimedia resource distributors, along with address and contact information, can be found in Appendix B (page 189). In addition, a list of supply houses is in Appendix C (page 193).

Suggested InterActive Physiology® modules are listed at the beginning of relevant exercises and included in Multimedia Resources in Appendix A. Students are enabled to understand, rather than memorize, difficult physiological concepts with these detailed interactive animations, puzzles, quizzes, and other tools. Eight major topic areas are covered: Muscular System; Nervous System I: The Neuron—The Action Potential; Nervous System II: Synaptic Potentials and Neurotransmitters; Cardiovascular System; Respiratory System; Urinary System; Fluids, Electrolytes, and Acid/Base Balance; and Endocrine System. Available on CD-ROM and at www.interactivephysiology.com.

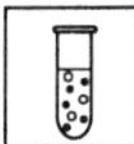

Each exercise includes directions for preparing needed solutions, indicated by the test tube icon.

Laboratory Safety

Always establish safety procedures for the laboratory. Students should be given a list of safety procedures at the beginning of each semester and should be asked to locate exits and safety equipment. Suggested procedures are on page v, along with a student acknowledgment form. These pages may be copied and given to the students. Signed student acknowledgment forms should be collected by the instructor once the safety procedures have been read and explained and the safety equipment has been located.

Special precautions must be taken for laboratories using body fluids. Students should use only their own fluids or those provided by the instructor. In many cases, suitable alternatives have been suggested. All reusable glass and plasticware should be soaked in 10% bleach solution for 2 hours and then washed with laboratory detergent and autoclaved if possible. Disposable items should be placed in an autoclave bag for 15 minutes at 121°C and 15 pounds of pressure to ensure sterility. After autoclaving, items may be discarded in any disposal facility.

Disposal of dissection materials and preservatives should be arranged according to state regulations. Be advised that regulations vary from state to state. Contact your state Department of Health or Environmental Protection Agency or their counterparts for advice. Keep in mind that many dissection specimens can be orderd in formaldehyde-free preservatives; however, even formaldehyde-free specimens may not be accepted by local landfill organizations.

Human Anatomy & Physiology Laboratory Safety Procedures

1. Upon entering the laboratory, locate exits, fire extinguisher, fire blanket, chemical shower, eye wash station, first aid kit, broken glass containers, and cleanup materials for spills.
2. Do not eat, drink, smoke, handle contact lenses, store food, and apply cosmetics or lip balm in the laboratory. Restrain long hair, loose clothing, and dangling jewelry.
3. Students who are pregnant, taking immunosuppressive drugs, or who have any other medical condition (e.g., diabetes, immunological defect) that might necessitate special precautions in the laboratory must inform the instructor immediately.
4. Wearing contact lenses in the laboratory is inadvisable because they do not provide eye protection and may trap material on the surface of the eye. If possible, wear regular eyeglasses instead.
5. Use safety glasses in all experiments involving liquids, aerosols, vapors, and gases.
6. Decontaminate work surfaces at the beginning and end of every laboratory period, using a commercially prepared disinfectant or 10% bleach solution. After labs involving dissection of preserved material, use hot soapy water or disinfectant.
7. Keep liquids away from the edge of the lab bench to help avoid spills. Liquids should be kept away from the edge of lab benches. Clean up spills of viable materials using disinfectant or 10% bleach solution.
8. Properly label glassware and slides.
9. Use mechanical pipetting devices; mouth pipetting is prohibited.
10. Wear disposable gloves when handling blood and other body fluids, mucous membranes, or nonintact skin, and/or when touching items or surfaces soiled with blood or other body fluids. Change gloves between procedures. Wash hands immediately after removing gloves. (Note: cover open cuts or scrapes with a sterile bandage before donning gloves.)
11. Place glassware and plasticware contaminated by blood and other body fluids in a disposable autoclave bag for decontamination by autoclaving or place them directly into a 10% bleach solution before reuse or disposal. Place disposable materials such as gloves, mouthpieces, swabs, and toothpicks that come into contact with body fluids into a disposable autoclave bag, and decontaminate before disposal.
12. To help prevent contamination by needle stick injuries, use only disposable needles and lancets. Do not bend needles and lancets. Needles and lancets should be placed promptly in a labeled puncture-resistant leakproof container and decontaminated, preferably by autoclaving.
13. Do not leave heat sources unattended.
14. Report all spills or accidents, no matter how minor, to the instructor.
15. Never work alone in the laboratory.
16. Remove protective clothing and wash hands before leaving the laboratory.

Laboratory Safety Acknowledgment Sheet

I hereby certify that I have read the safety recommendations provided for the laboratory and have located all of the safety equipment listed in Safety Procedure Number 1 of these procedures.

Student's Name ______________________________

Course ______________________________ Date ____________

Instructor's Name ______________________________

Adapted from:

Biosafety in Microbiological and Biomedical Laboratories. 1988. U.S. Government Printing Office, Washington, D.C. 20402.

Centers for Disease Control. 1989. "Guidelines for Prevention of Transmission of Human Immunodeficiency Virus and the Hepatitis B Virus to Health-Care and Public-Safety Workers." *MMWR*: 38 (S6).

———. 1987. "Recommendations for Prevention of HIV Transmission in Health-Care Settings." *MMWR*: 36 (2s).

Johnson, Ted, and Christine Case. 1992. *Laboratory Experiments in Microbiology, Brief Version*, Third Edition. Redwood City, CA: Benjamin/Cummings Publishing Co.

School Science Laboratories: A Guide to Some Hazardous Substances. 1984. U.S. Consumer Product Safety Commission. Washington, D.C. 20207.

U.S. Department of Health and Human Services Centers for Disease Control and Prevention and National Institutes for Health, Fourth Edition. May 1999. U.S. Government Printing Office. Washington, D.C. http://www.cdc.gov.od/ohs/manual/labsfty.htm.

Contents

Part One: Exercises 1

Part Two: Answers to Review Sheets 59

Appendices 181

exercise 1

The Language of Anatomy

If time is a problem, most of this exercise can be done as an out-of-class assignment.

Time Allotment: (in lab): 1/2 hour.

Multimedia Resources: See Appendix A for a list of multimedia resource distributors.
Systems Working Together (WNS, 15 minutes, VHS)
The Human Body: The Ultimate Machine (CBS, 22 minutes, VHS)
The Incredible Human Machine (CBS, 60 minutes, VHS)

Interactive Atlas of Human Anatomy (ICON, CD-ROM)

Advance Preparation

1. Set out human torso models and have articulated skeletons available.
2. Obtain three preserved kidneys (sheep kidneys work well) and three bananas. Cut one of each in transverse section, one in longitudinal section (usually a sagittal section), and leave one uncut. Label the kidneys and put them in a demonstration area. You may wish to add a fourth kidney to demonstrate a frontal section.
3. The day before the lab, prepare gelatin or Jell-O® using slightly less water than is called for and cook the spaghetti until it is al dente. Pour the gelatin into several small molds and drop several spaghetti strands into each mold. Refrigerate until lab time.
4. Set out gelatin spaghetti molds and scalpel.

Comments and Pitfalls

1. Students will probably have the most trouble understanding proximal and distal; other than that there should be few problems.

Answers to Questions

Activity 3: Practicing Using Correct Anatomical Terminology (p. 3)

1. The wrist is *proximal* to the hand.
2. The trachea (windpipe) is *anterior* or *ventral* to the spine.
3. The brain is *superior* or *cephalad* to the spinal cord.
4. The kidneys are *inferior* or *caudal* to the liver.
5. The nose is *medial* to the cheekbones.
6. The chest is *superior* to the abdomen.

exercise

Organ Systems Overview

2

Time Allotment: 1 1/2 hours (rat dissection—1 hour; human torso model—1/2 hour).

Multimedia Resources: See Appendix A for a list of multimedia resource distributors.
Homeostasis (FHS, 20 minutes, VHS)
Homeostasis: The Body in Balance (HRM, IM, 26 minutes, VHS)
The Human Body: The Ultimate Machine (CBS, 27 minutes, VHS)
The Incredible Human Machine (CBS, 60 minutes, VHS)
Organ Systems Working Together (WNS, 14 minutes, VHS)

Advance Preparation

1. Make arrangements for appropriate storage and disposal of dissection materials. Check with the Department of Health or the Department of Environmental Protection, or their counterparts, for state regulations.
2. Designate a disposal container for organic debris, set up a dishwashing area with hot soapy water and sponges, and provide lab disinfectant such as Wavicide-01 (Carolina) for washing down the lab benches.
3. Set out safety glasses and disposable gloves for dissection of freshly killed animals (to protect students from parasites) and for dissection of preserved animals.
4. Decide on the number of students in each dissecting group (a maximum of four is suggested, two is probably best). Each dissecting group should have a dissecting pan, dissecting pins, scissors, blunt probe, forceps, twine, and a preserved or freshly killed rat.
5. Preserved rats are more convenient to use unless small mammal facilites are available. If live rats are used, they may be killed a half hour or so prior to the lab by administering an overdose of ether or chloroform. To do this, remove each rat from its cage and hold it firmly by the skin at the back of its neck. Put the rat in a container with cotton soaked in ether or chloroform. Seal the jar tightly and wait until the rat ceases to breathe.
6. Set out human torso models and a predissected rat.

Comments and Pitfalls

1. Students may be overly enthusiastic when using the scalpel and cut away organs they are supposed to locate and identify. Have blunt probes available as the major dissecting tool and suggest that the scalpel be used to cut only when everyone in the group agrees that the cut is correct.

2. Be sure the lab is well ventilated, and encourage students to take fresh air breaks if the preservative fumes are strong. If the dissection animal will be used only once, it can be rinsed to remove most of the excess preservative.
3. Organic debris may end up in the sinks, clogging the drains. Remind the students to dispose of all dissection materials in the designated container.

Answers to Questions

Activity 6: Examining the Human Torso Model (pp. 13–14)

Digestive: *esophagus, liver, stomach, pancreas, small intestine, large intestine (including rectum)*

Urinary: *kidneys, ureters, bladder*

Cardiovascular: *heart, descending aorta, inferior vena cava*

Endocrine: *thyroid gland, pancreas, adrenal gland*

Reproductive: *none*

Respiratory: *lungs, bronchi, trachea*

Lymphatic: *spleen*

Nervous: *brain, spinal cord, medulla of adrenal gland*

The Cell—Anatomy and Division

exercise

3

The Anatomy of the Composite Cell section can be given as an out-of-class assignment to save time. This might be necessary if audiovisual material is used.

Time Allotment: 2 hours.

Multimedia Resources: See Appendix A for a list of multimedia resource distributors.
A Journey Through the Cell (FHS, 2-part series, 25 minutes each, VHS, DVD)
An Introduction to the Living Cell (CBS, 30 minutes, VHS)
Inside the Living Cell (WNS, set of 5, VHS)
Mitosis and Meiosis (UL, 23 minutes, VHS)

The Cell: Structure, Function, and Process (HRM, CD-ROM)
Inside the Cell (CE, CD-ROM)
Mitosis (CE, CD-ROM)

Advance Preparation

1. Set out slides (one per student) of simple squamous epithelium, teased smooth muscle, human blood cell smear, sperm, and whitefish blastulae. Students will also need lens paper, lens cleaning solution, immersion oil, and compound microscopes.
2. Set out a model or a lab chart of a composite cell, and models of mitotic stages.
3. Set out pipe cleaners and chalk.
4. If available, arrange a viewing area for the mitosis video.

Comments and Pitfalls

1. Observing differences and similarities in cell structure often gives students trouble, as many of them have never seen any cells other than epithelial cells. Slides or pictures of these cell types might help.

Answers to Questions

Activity 5: Observing Differences and Similarities in Cell Structure (pp. 17–18)

3. Simple squamous epithelial cells are relatively large and irregularly ("fried egg") shaped. Smooth muscle cells are also relatively large, but are long and spindle shaped. Red blood cells and sperm are both examples of small cells. Red blood cells appear round, while sperm cells are streamlined with long flagella.

   Cell shape is often directly related to function. Epithelial cells fit tightly together and cover large areas. Elongated muscle cells are capable of shortening during contraction. The red blood cells are small enough to fit through capillaries, and are actually biconcave in shape, which makes them flexible and increases surface area (not obvious to the students at this point). Sperm cells' streamlined shape and flagella are directly related to efficient locomotion.

   The sperm cells have visible projections (flagella), which are necessary for sperm motility. The function of sperm is to travel through the female reproductive system to reach the ovum in the uterine tubes. This requires motility.

   None of the cells lack a plasma membrane. Mature red blood cells have no nucleus. Nucleoli will probably be clearly visible in the epithelial cells, and possibly visible in the other nuclei.

   No. Identifiable organelles are not visible in most of these cells. Filaments may be visible in the smooth muscle preparations. The details of organelle structure are usually below the limit of resolution of the light microscope. Unless special stains are used, there is no way to see or distinguish the organelles at this level.

Cell Membrane Transport Mechanisms

exercise

4

This exercise has many parts to it. If students have had an introductory cell biology course, much of it should be review.

Time Allotment:
Diffusion of dye through agar gel—90 minutes
Diffusion through nonliving membranes—120 minutes
Diffusion through living membranes—25 minutes
Filtration—10 minutes

Observations for diffusion through living membranes and filtration can be done while waiting for the results of the other experiments.

Multimedia Resources: See Appendix A for a list of multimedia resource distributors.
A Journey Through the Cell (FHS, 2-part series, 25 minutes each, VHS, DVD)
Mitosis and Meiosis (UL, 23 minutes, VHS)
The Outer Envelope (WNS, 15 minutes, VHS)

Mitosis (CE, CD-ROM)

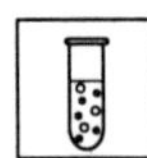

Solutions:
Agar Gel, 1.5%
Weigh out 15 grams of dried agar. Slowly add 1 liter of distilled water while heating. Bring slowly to a boil, stirring constantly until the agar dissolves. For immediate use, allow the agar to cool to about 45°C. Pour into petri dishes to solidify. Refrigerate in an inverted position. If the plates are to be kept for a longer time (more than one day), autoclave the agar solution in the flask, pour into sterile petri plates, allow the agar to solidify, invert the plates, and store in a refrigerator.

Benedict's Solution
- 173.0 grams sodium citrate
- 100.0 grams sodium carbonate, anhydrous
- 17.3 grams cupric sulfate (pure crystalline)

Add the citrate and carbonate salts to 700–800 milliliters distilled water and heat to dissolve. Add the cupric sulfate to 100 milliliters distilled water and heat to dissolve. Cool the solutions and then combine. Add distilled water to make 1 liter of solution.

Glucose, 40%
For each 100 milliliters of solution, weigh out 40 grams of glucose and bring to 100 milliliters with distilled water. It may be necessary to heat the mixture to get the glucose into solution. Refrigerate when not in use.

Lugol's Iodine (IKI)
• 20 grams potassium iodide
• 4 grams iodine crystals
Dissolve potassium iodide in 1 liter distilled water. Add the iodine crystals and stir to dissolve. Store in dark bottles.

Physiologic Saline (Mammalian, 0.9%)
Weigh out 9 grams of NaCl. Add water to a final volume of 1 liter. Make fresh immediately prior to experiment.

Silver Nitrate (2.9 or 3%)
Weigh out 2.9 grams (for 2.9%) or 3 grams (for 3%) of silver nitrate. **Use caution, this is an oxidizing substance**. Add distilled water to make 100 milliliters of solution. Store in light-resistant bottles. Make fresh for each use.

Sodium Chloride (NaCl), 10%
For each 100 milliliters of solution, weigh out 10 grams of NaCl and bring to 100 milliliters with distilled water. It may be necessary to heat the mixture to get the NaCl into solution.

Sodium Chloride (NaCl), 1.5% saline
Weigh out 1.5 grams NaCl. Add distilled water to a final volume of 100 milliliters.

Sucrose, 40% (with Congo Red Dye)
For each 100 milliliters of solution, weigh out 40 grams of sucrose and bring to 100 milliliters with distilled water. Add Congo red dye as necessary to color the solution red. It may be necessary to heat the solution to get the sucrose into the solution. Refrigerate when not in use.

Advance Preparation

Note: This lab has many components. Either clearly designate supply areas for each part of the lab, or provide each lab group with its own set of supplies at the outset. The supplies for each part of the exercise are listed separately in case sections of the exercise are omitted. Some equipment is common to several parts of the lab.

1. Set out slides and coverslips. Have compound microscopes available.
2. *Diffusion of Dye Through Agar Gel.* Set out 0.1M or 3.5% methylene blue solution (Carolina) and 0.1M or 1.6% potassium permanganate solution (Carolina), 1.5% agar plates (12 milliliters of 1.5% agar per plate, one per group), medicine droppers, and millimeter rulers.
3. *Diffusion Through Nonliving Membranes.* For each group, set out four dialysis sacs (Ward's) or 10-centimeter lengths of dialysis tubing (Carolina), five 250-milliliter beakers, a wax marking pencil, 750 milliliters of distilled water, 10 milliliters of *10%*

NaCl solution, 10 milliliters of *Congo red dye in 40% sucrose*, 150 milliliters of *40% glucose solution*, dropper bottles of *Benedict's solution* (Carolina, or see above), *silver nitrate*, and *Lugol's iodine* (Carolina, or see above), four test tubes, a test tube rack, test tube holder, small graduated cylinder, a small funnel, hot plate, and balance. Dialysis sacs can be prepared from cut sections of dialysis tubing. Soak dialysis tubing in a beaker of water for about 15 minutes. Once dialysis tubing has been soaked, open it by rubbing it between the thumb and forefinger until the tubing material separates. Tie the ends with fine twine or close with dialysis tubing closures (Carolina). Small Hefty® "alligator" sandwich bags can also be used to make dialysis bags.

4. *Diffusion Through Living Membranes*. To prepare the solutions, add 5–10 drops of blood to test tubes containing: physiological saline (specimen 1), *1.5% NaCl* (specimen 2), and distilled water (specimen 3). Drop 1–2 drop(s) of each solution on three separate slides and cover with coverslips. Set up demonstration area with compound microscopes and red blood cell suspensions.
5. *Filtration*. For each group, set out a funnel, filter paper, a flask, a dropper bottle of *Lugol's iodine*, 50 milliliters of a solution of uncooked starch, powdered charcoal, and copper sulfate.

Comments and Pitfalls

1. Caution students to keep careful track of time during the diffusion experiments. Lab timers might help. Suggestions for variables include T and different concentrations of solutions.
2. Dialysis sacs may leak. Check to see that they are tightly sealed.
3. You may substitute Clinitest tablets for Benedict's solution.
4. Silver nitrate will stain and possibly damage clothing. Warn students to be careful.
5. Note that the *40% glucose solution* used in sac 1 of the osmosis experiment is not iso-osmotic to the *10% NaCl solution* in sac 3, so caution students about the types of conclusions they may draw from this experiment. Also, sometimes no glucose will be present in the beaker at the end of the hour. You may need to extend the time for this part of the experiment.
6. Emphasize the importance of labeling test tubes and slides.
7. Red blood cells in physiologic saline may begin to crenate as the slide begins to dry out. Encourage students to make their observations quickly. If there is still trouble with crenation, use a slightly hypotonic saline solution.

Answers to Questions

Activity 1: Observing Diffusion of Dye Through Agar Gel (pp. 22–23)

5. Potassium permanganate (MW 158) diffused more rapidly than methylene blue (MW 320). The smaller the molecular weight, the faster the rate of diffusion. The dye molecules moved because they possess kinetic energy.

   Heating the samples will increase the rate of diffusion for both samples by increasing the kinetic energy of the molecules. The greater the kinetic energy the faster the diffusion.

Activity 2: Observing Diffusion Through Nonliving Membranes (pp. 23–25)

4. After 1 hour, sac 1 (originally containing 40% glucose) should have gained weight. Water is moving into the sac by osmosis. Glucose is still present in the sac, and a small amount of glucose may also be present in the beaker. If the Benedict's test is positive, glucose was able to pass through the dialysis membrane.
5. There should be no net weight change in sac 2. Since the concentrations of glucose and water are the same on both sides of the membrane, there is no net movement of water or glucose.
6. Sac 3 will increase in weight, perhaps only by a small amount. There has been a net movement of water into the sac and the weight of the water was not completely offset by the movement of the NaCl out of the sac. The solution in beaker 3 reacts with silver nitrate, indicating the presence of chloride in the beaker. Net dialysis of NaCl occurred.
7. There should be an increase in weight in sac 4. The water color did not turn pink; the dye was not able to diffuse out of the sac.

   The Benedict's test for sugar was negative. Sucrose did not diffuse from the sac to the beaker. The dye and sucrose molecules are too large to diffuse through the pores in the membrane or their rate of diffusion is too slow given the allowed time.
8. Net osmosis occurred in situations 1 and 4; net simple diffusion occurred in 1 and 3.

   Water molecules are very small, and move quickly down a concentration gradient. Na^+ and Cl^- in solution behave like slightly larger molecules, but are smaller than glucose molecules, which move slowly, if at all, through the dialysis tubing. (See item 5 in Comments and Pitfalls.) Note: Students may only be able to conclude that Na^+ and Cl^- in solution and water molecules are small, and glucose and Congo red dye molecules are larger, or that Na^+ and Cl^- in solution and water and glucose molecules are smaller than sucrose molecules.

   The dialysis sac is often compared to the plasma membrane of the cell.

Activity 3: Investigating Diffusion Through Living Membranes (p. 25)

2. Physiological saline is isotonic, therefore the suspended cells will not change shape.
3. The cells begin to shrink and develop a multipointed star shape.
4. When distilled water is added the cells should begin to revert to their normal shape. Eventually they begin to look very bloated, and finally begin to disappear as their membranes burst open.

   Since the intact cells are no longer present in solution following hemolysis, test tube C solution appears clear.

Activity 4: Observing the Process of Filtration (pp. 25–26)

3. Copper sulfate passes through the filter paper; starch and charcoal are retained.

   The filter paper represents a filtration barrier like the capillary endothelium, basement membrane, and podocytes of the renal corpuscle.

   The filtration rate was greatest during the first interval due to greater hydrostatic pressure in the funnel.

   The size of these molecules determines their ability to pass through the filter paper.

exercise

Classification of Tissues

5

Time Allotment: 2 hours.

Multimedia Resources: See Appendix A for a list of multimedia resource distributors.
Basic Human Histology (CBS, microslide sets of eight related 35-mm slides)
Histology Slides for Life Science (BC, 35-mm slides)

Histology Videotape Series (UL, 26-part series, VHS)

Eroschenko's Interactive Histology (UL, CD-ROM)
PhysioEX™: Exercise 6B (BC, CD-ROM)
Ward's Histology Collection (WNS, CD-ROM)

Advance Preparation

1. Set out prepared slides of simple squamous, simple cuboidal, simple columnar, stratified squamous (nonkeratinized), pseudostratified ciliated columnar, and transitional epithelium.
2. Set out prepared slides of adipose, areolar, reticular, and dense regular (tendon) connective tissue; of hyaline cartilage; and of bone (cross section).
3. Set out prepared slides of skeletal, cardiac, and smooth muscle (longitudinal sections).
4. Set out prepared slide of nervous tissue (spinal cord smear).
5. Set out lens paper and lens cleaning solution. Have compound microscopes available.

Comments and Pitfalls

1. Slides of the lung are suggested for simple squamous epithelium and slides of the kidney are suggested for simple cuboidal epithelium.
2. The dense fibrous regular connective tissue slide is sometimes labeled "white fibrous tissue."
3. Students may have trouble locating the appropriate tissue on slides with multiple tissue types. Encourage them to consult lab manual Figures 5.2–5.5, the Histology Atlas on pp. 363–373, and each other for help.
4. A television camera with a microscope adapter and monitor is very useful in this lab. By watching the monitor, students can observe the instructor locating the correct area of

tissue on the slide (see item 3 in Comments and Pitfalls). It also makes it easier to answer student questions and share particularly good slides with the class.

5. Constructing a concept map or dichotomous key of tissue types

   Constructing the map or key helps students clarify differences and similarities between tissues based on observations. It provides practice with observation, logical thinking, and grouping—skills that can be applied to material throughout the course.

   The map should also make it much easier for students to identify tissues on the slides.

   a. To construct a map of the tissues, prepare a series of questions that will separate the tissue types in some logical way. Each question should have only "yes" and "no" as possible answers. You will know that a branch of the map is complete when you have separated a group of tissues into single choices. Figure 5.6 in the lab manual suggests a way to begin.

   b. Read through Exercise 5 in the lab manual and note the general characteristics of epithelial, connective, muscle, and nervous tissue. Base the map on things you can *observe* with the microscope. For example, epithelial tissue is mitotic, but this is not something you can observe easily on the slides or pictures.

      Separate out the epithelial tissue pictures. Notice that each of these tissues has a *free edge*.

      A good first question might be, "Is there a free edge?" Note that there are only two possible answers—yes and no. This should separate all of the epithelial tissues from connective, muscle, and nervous tissue.

   c. Continue asking yes and no questions about the epithelial tissues until you have separated each of the epithelial tissue types into a separate category.

   d. Now turn your attention to the pictures in the other ("no") pile. Work on a set of questions that will separate each of these tissue types into a separate category.

exercise

6

The Skin (Integumentary System)

Time Allotment: 1 1/2 hours.

Multimedia Resources: See Appendix A for a list of multimedia resource distributors.

How the Body Works: Skin, Bones, and Muscles (AIMS, NIMCO, 19 minutes, VHS, DVD)
The Senses: Skin Deep (FHS, 26 minutes, VHS)
Skin (FHS, 20 minutes, VHS, DVD)
The Skin (NIMCO, 30 minutes, VHS)

How the Body Works: Skin, Bones, and Muscles (AIMS, NIMCO, CD-ROM, DVD)

Solution:
Lugol's Iodine (IKI)
• 20 grams potassium iodide
• 4 grams iodine crystals
Dissolve potassium iodide in 1 liter distilled water. Add the iodine crystals and stir to dissolve. Store in dark bottles.

Advance Preparation

1. Set out models of the skin, prepared slides of human skin with hair follicles, lens paper, and lens cleaning solution. Have compound microscopes available.
2. Terminology for layers of the epidermis differs from text to text. Decide on the terminology to be used, and inform the students at the onset of the laboratory session if there is a discrepancy between the laboratory manual and the text.
3. Set out 20# bond paper ruled to mark off cm^2 areas, Betadine swabs or *Lugol's iodine* (Carolina, or see above), cotton swabs, scissors, and adhesive tape. Also set out glass plates, calipers, coins, and felt-tipped markers.

Comments and Pitfalls

1. Students may have difficulty finding the arrector pili muscles and sweat glands. Some students will confuse the fibers of the dermis (dense fibrous irregular connective tissue) with smooth muscle.

2. Slides of the palmar or plantar skin provide a good contrast to slides of skin with hair follicles.
3. Cut the squares of bond paper before the lab begins to save time.

Answers to Questions

Activity 2: Visualizing Changes in Skin Color Due to Continuous External Pressure (p. 43)

The color of compressed skin will be pale.

The reason for the color change is obstruction to the capillary flow to the area.

If pressure is maintained long enough, the skin under pressure will obtain a bluish color. This condition is called cyanosis and is due to the presence of deoxygenated blood because blood flow has been blocked.

Activity 4: Determining the Two-Point Threshold (p. 44)

3. It would be reasonable to predict that fingertips and lips would have the greatest density of touch receptors.

Activity 5: Testing Tactile Localization (p. 44)

2. The ability to locate the stimulus should not improve with repeated trials because the receptor density remains unchanged.
4. Fingertips have the smallest error of localization because they have the highest density of touch receptors per area. Therefore, they are the most sensitive.

Activity 6: Demonstrating Adaptation of Touch Receptors (p. 45)

3. The pressure sensation returns when coins are added to the stack. The same receptors are probably being used. Generator potentials are graded and stronger stimuli produce larger potentials and thus increased frequency of nerve impulses.

Activity 7: Plotting the Distribution of Sweat Glands (pp. 45–46)

5. The palm has a greater density of sweat glands than the forearm.

Activity 8: Examining a Skin Slide (p. 47)

The stratified squamous epithelium of the skin is comprised of several recognizable layers, the outermost of which are keratinized.

Both types of epithelia are protective, but the skin epithelium also protects against water loss to the external environment, UV damage, and chemical damage in addition to protecting against mechanical damage and bacterial invasion.

exercise

Overview of the Skeleton

7

Time Allotment: 45 minutes.

Multimedia Resources: See Appendix A for a list of multimedia resource distributors.
How the Body Works: Skin, Bones, and Muscles (AIMS, NIMCO, 19 minutes, VHS, DVD)
Human Skeletal System (IM, 23 minutes, VHS)
Muscle and Bone (NIMCO, 30 minutes, VHS)
Our Flexible Frame (WNS, 20 minutes, VHS)
Skeletal System: The Infrastructure (FHS, 25 minutes, VHS, DVD)
The Skeletal System (WNS, 15 minutes, VHS)
Skeleton: An Introduction (UL, 46 minutes, VHS)

How the Body Works: Skin, Bones, and Muscles (AIMS, NIMCO, CD-ROM, DVD)

Solution:
Nitric Acid (HNO_3), 10%
Put 80 milliliters of distilled water into a graduated cylinder.
Carefully add 14.3 milliliters of 69–71% nitric acid.
Add water to a final volume of 100 milliliters;
or put 50 milliliters of distilled water in a graduated cylinder.
Carefully add 10 milliliters of 69–71% nitric acid.
Add water to a final volume of 70 milliliters.

Advance Preparation

1. If you have a local source, arrange to have a long bone sawed longitudinally. Keep refrigerated or frozen until used. Preserved, sawed longbones can be used instead. Provide disposable gloves at the demonstration area.
2. Bake some long bones (chicken or turkey bones work well) at 250°F for 2 hours or until they are brittle and snap or crumble easily. Prepare these the day before lab observations are to take place.
3. Soak some long bones in *10% nitric acid* or vinegar until flexible. Overnight soaking is usually sufficient for the nitric acid; vinegar will take longer. Prepare well in advance.

4. Prepare numbered samples of long, short, flat, and irregular bones. These can be set out at a station in the lab where students can work on identification.
5. Put out a prepared slide of ground bone (cross section), lens paper, and lens cleaning solution. Have compound microscopes available.
6. Set out models of the microscopic structure of bone and an articulated skeleton.

Comments and Pitfalls

1. Students initially may have some trouble classifying bones by shape; other than that, this lab should cause no problems.
2. Emphasize that all long bones have a long axis, but some long bones are much shorter than others! Long bones include most of the bones of the upper and lower limbs (humerus, radius, ulna, femur, tibia, fibula, metacarpals, metatarsals, phalanges). Short bones include the carpals and the tarsals. Flat bones are thin and include the bones of the roof of the cranial cavity, sternum, scapula, and ribs. Irregular bones include some skull bones, the vertebrae, and possibly bones of the pelvic girdle. Bones included in each of these categories vary from author to author.
3. HCl may be substituted for HNO_3.

Answers to Questions

Activity 3: Comparing the Relative Contributions of Bone Salts and Collagen Fibers in Bone Matrix (p. 53)

The cup will support the heavy book. This is comparable to the compression strength provided by bone salts to the bone.

The leather belt will not break. This is comparable to the tensile strength of bone provided by bone collagen fibers.

Activity 4: Examining the Effects of Heat and Nitric Acid on Bones (pp. 53–54)

The treated bones still have the same general shape as the untreated bones, although the acid-soaked bone may appear more fibrous.

The heated bone is very brittle and responds to gentle pressure by breaking.

The acid-treated bone is very flexible.

The acid appears to remove the calcium salts from the bone.

Heating dries out the organic matrix.

The acid-treated bone most closely resembles the bones of a child with rickets.

exercise

The Axial Skeleton

8

Time Allotment: 2½ hours.

Multimedia Resources: See Appendix A for a list of multimedia resource distributors.
The Human Skeletal System (IM, 23 minutes, VHS)
Skeletal System: The Infrastructure (FHS, 27 minutes, VHS, DVD)
The Skull Anatomy Series (UL, 9-part series, VHS)
The Thoracic Skeleton (UL, 18 minutes, VHS)

Interactive Skeleton: Sports and Kinetic (LP, CD-ROM)

Advance Preparation

1. Set out one intact adult skull per group.
2. Set out an isolated fetal skull in a demonstration area, unless enough are available for each group.
3. Set out labeled samples of disarticulated vertebrae, an articulated spinal column, a disarticulated skull, and a Beauchene skull.
4. Have articulated skeletons available. There should be a minimum of two, one male and one female.
5. Display X rays of individuals with scoliosis, kyphosis, and lordosis, if available. Students are often willing to bring in X rays for the class to use if none are available.
6. Set out blunt probes or unsharpened pencils with erasers for the students to use while studying the bones. Caution them against marking the bones with pencils or markers.

Comments and Pitfalls

1. Students may have some trouble with the numerous foramina of the skull. You may wish to have them locate all of the foramina at this time, but hold them responsible for identifying a smaller number.
2. The ethmoid bone may cause some problems, especially if the skulls are old and the conchae have begun to crumble. The disarticulated and Beauchene skulls will come in handy here.
3. There is the occasional student who asks whether males have one less rib than females. A trip to the articulated skeletons provides the answer: no.

Answers to Questions

Activity 2: Examining a Fetal Skull (p. 60)

1. Yes, the fetal and adult skulls have the same bones, although the fetal frontal bone is bipartite as opposed to the single frontal bone seen in the adult skull.

   The fetal face is foreshortened and overshadowed by the cranium; the maxillae and mandible are very tiny.

   In the adult skull the cranium is proportionately smaller and the facial skeleton proportionately larger.

Activity 4: Examining Spinal Curvatures (p. 61)

2. When the fibrous disc is properly positioned, the spinal cord and peripheral nerves are not impaired in any way. If the disc is removed, the intervertebral foramina are reduced in size, and might pinch the nerves exiting at that level.

   Slipped discs often put pressure on spinal nerves, causing pain and/or loss of feeling.

exercise

The Appendicular Skeleton

9

Time Allotment: 2 hours.

Multimedia Resources: See Appendix A for a list of multimedia resource distributors.

Anatomy of a Runner (Structure and Function of the Lower Limb) (UL, 38 minutes, VHS)
Anatomy of the Hand (FHS, 14 minutes, VHS, DVD)
Anatomy of the Shoulder (FHS, 18 minutes, VHS, DVD)
Bones and Joints (FHS, 20 minutes, VHS, DVD)
Gluteal Region and Hip Joint (UL, 18 minutes, VHS)
Knee Joint (UL, 16 minutes, VHS)

Interactive Foot and Ankle (LP, CD-ROM)
Interactive Shoulder (LP, CD-ROM)

Advance Preparation

1. Have articulated skeletons (male and female) available.
2. Set out disarticulated skeletons. One per group of 3–4 students is ideal.
3. Set out male and female articulated pelves in a demonstration area.
4. Set out blunt probes or unsharpened pencils with erasers for use during bone identification.
5. Set out X rays of bones of the appendicular skeleton.
6. Set out six numbered bones concealed in Kraft paper or muslin bags.

Comments and Pitfalls

1. Students may have trouble distinguishing between right and left samples of bones. Remind them to review the bone markings before checking the articulated skeleton.
2. Stress the importance of bony landmarks for muscle location and identification.

exercise

Joints and Body Movements

10

Time Allotment: 1 hour.

Multimedia Resources: See Appendix A for a list of multimedia resource distributors.
Anatomy of a Runner (Structure and Function of the Lower Limb) (UL, 38 minutes, VHS)
Bones and Joints (FHS, 20 minutes, VHS, DVD)
Gluteal Region and Hip Joint (UL, 18 minutes, VHS)
Knee Joint (UL, 16 minutes, VHS, DVD)
Movement at Joints of the Body (FHS, 40 minutes, VHS, DVD)
Moving Parts (FHS, 26 minutes, VHS)
The Skeleton: Types of Articulation (UL, 16 minutes, VHS)

Advance Preparation

1. If you have a local source, obtain a sagittally sawed, fresh diarthrotic beef joint from a butcher or meat packing company. Refrigerate or freeze until use. Preserved joints could be used instead. Have disposable gloves available.
2. Have available the articulated skeleton and isolated skull.
3. Set out any available anatomical charts of joint types, models of joint types, etc., that are available.
4. Display any available X rays of normal and arthritic joints.
5. There are several methods of joint classification. If your text and the lab manual use different systems, decide on the preferred system for your course.
6. Have water balloons and clamps available.

Comments and Pitfalls

1. Some students may have trouble interpreting the movements in Figure 10.4. It may help to have the students perform all of these movements together during lab.
2. Students may be confused by movement at the shoulder joint. Flexion occurs when the arm is moved forward and upward, and extension returns the arm to the anatomical position.

Answers to Questions

Activity 4: Demonstrating the Importance of Friction-Reducing Structures (pp. 79–80)

3. The fluid-filled sac greatly reduces the friction between the two surfaces. The water balloon represents a synovial cavity, bursae, or tendon sheaths. The fists represent two articulating bones on opposite sides of a synovial cavity. They may also represent muscles, tendons, or ligaments in the case of bursae and tendon sheaths.

exercise

Microscopic Anatomy and Organization of Skeletal Muscle

11

Time Allotment: 2 hours.

Multimedia Resources: See Appendix A for a list of multimedia resource distributors. See Exercise 5 for histology listings.
Human Musculature Videotape (BC, 23 minutes, VHS)
Muscles (FHS, 20 minutes, VHS, DVD)
Muscles and Joints: Muscle Power (FHS, 26 minutes, VHS, DVD)
The Skeletal and Muscle Systems (UL, 24 minutes, VHS)

InterActive Physiology® 8-System Suite–Muscular System (BC, CD-ROM or www.interactivephysiology.com)
Anatomy Review: Skeletal Muscle Tissue
The Neuromuscular Junction
Sliding Filament Theory
Muscle Metabolism
Contraction of Motor Units
Contraction of Whole Muscle

Advance Preparation

1. Set out prepared slides of skeletal muscle (longitudinal and cross sections), and slides showing neuromuscular junctions. Set out lens paper and lens cleaning solution. Have compound microscopes available.
2. Set out any available models of skeletal muscle cells and neuromuscular junctions.

Comments and Pitfalls

1. Students may have difficulty observing the muscle banding pattern. This is usually because the light intensity is set too high and the iris diaphragm is not closed down.
2. Emphasize the importance of understanding the organization and terminology of muscle structure. The organization and terminology of the nerves are very similar.

Advance Preparation—ATP Muscle Kit

1. Order the ATP muscle kits (Carolina) to be delivered no more than seven days before the lab. One kit provides generously for eight students. Extra vials of the chemical solutions can be ordered separately (Carolina) and will reduce waiting time. Just before the lab begins, cut the muscle bundles into 2-centimeter lengths and place in a petri dish in the accompanying glycerol.
2. Glass dissecting needles can be made easily from glass stirring rods. Use a Bunsen burner with a flame spreader attachment. Holding a stirring rod with oven mitts, heat the center while turning the rod until the flamed area glows orange. Pull the ends gently but firmly apart until the glass separates. With practice, fine-tipped needles can be made.

Comments and Pitfalls

1. Students may have great difficulty separating the muscle bundles into individual fibers. Often two or three fibers remain together and it is the best they can do.
2. Remind the students to keep the fibers in a pool of glycerol to prevent them from drying out.
3. Sometimes the fibers curl as they contract. Caution the students to measure the uncurled length of the fiber.
4. Occasionally there is great variability in the results (probably due to technical errors). Try rinsing the slides and glass needles in distilled water before use. This is a good exercise to collect class data and have the students compare individual results with the class results. You can discuss the importance of controlled experiments and repeated trials.

Answers to Questions

Activity 2: Observing Muscle Fiber Contraction (pp. 85–86)

5. Generally there is little or no contraction with ATP alone. There is no contraction with the salt solutions alone. Maximum contraction occurs in the presence of ATP and the proper concentrations of potassium and magnesium ions.

exercise

Gross Anatomy of the Muscular System

12

Time Allotment: 2–3 hours in lab plus time outside of lab.

Multimedia Resources: See Appendix A for a list of multimedia resource distributors.
Anatomy of a Runner (Structure and Function of the Lower Limb) (UL, 38 minutes, VHS)
Abdomen and Pelvis (UL, 16 minutes, VHS)
Human Musculature Videotape (BC, 23 minutes, VHS)
Lower Extremity (UL, WNS, 28 minutes, VHS)
Major Skeletal Muscles and their Actions (UL, 19 minutes, VHS)
Muscles (FHS, 20 minutes, VHS, DVD)
The Skeletal and Muscular Systems (UL, 24 minutes, VHS)

Advance Preparation

1. Set out models of the human torso and upper and lower limbs. It helps to have the muscles labeled on some of the models. Have model keys available.
2. Set out anatomical charts of human musculature.
3. Set out tubes of body (or face) paint and 1-inch wide brushes.

exercise

Neuron Anatomy & Physiology

13

Time Allotment: 1 hour.

Multimedia Resources: See Appendix A for a list of multimedia resource distributors. See Exercise 5 for histology listings.

Brain and Nervous System: Your Information Superhighway (FHS, 25 minutes, VHS, DVD)
The Central Nervous System and Brain (IM, 29 minutes, VHS)
The Nature of the Nerve Impulse (FHS, 15 minutes, VHS, DVD)
Nerves and Nerve Cells (NIMCO, 28 minutes, VHS)
The Nervous System: Nerves at Work (FHS, 27 minutes, VHS)

InterActive Physiology® 8-System Suite–Nervous System I and II (BC, CD-ROM or www.interactivephysiology.com)

Nervous I	**Nervous II**
Orientation	Orientation
Anatomy Review	Anatomy Review
Ion Channels	Ion Channels
The Membrane Potential	Synaptic Potentials and Cellular Integration
The Action Potential	Synaptic Transmission

Advance Preparation

1. Set out slides of ox spinal cord smear, teased myelinated nerve fibers, and nerve cross section.
2. Set out lens paper, immersion oil, and lens cleaning solution. Have compound microscopes available.
3. Set out models of neurons, if available.
4. Set up videotape viewing area.

Comments and Pitfalls

1. Students may focus on the wrong cells. Encourage them to use the Histology Atlas on pp. 363–373, and help each other.

Answers to Questions

Activity 1: Identifying Parts of a Neuron (pp. 105–106)

3. The nodes are at regular intervals. Action potentials will occur at regular intervals along the axon as local currents open voltage-gated sodium channels.

Gross Anatomy of the Brain and Cranial Nerves

exercise

14

Time Allotment: 2 hours.

Multimedia Resources: See Appendix A for a list of multimedia resource distributors.

Anatomy of the Human Brain (FHS, 35 minutes, VHS, DVD)
Animated Neuroscience and the Action of Nicotine, Cocaine, and Marijuana in the Brain (FHS, 25 minutes, VHS, DVD)
The Brain (FHS, 20 minutes, VHS)
The Brain (NIMCO, 30 minutes, VHS)
Brain and Nervous System: Your Information Superhighway (FHS, 25 minutes, VHS, DVD)
The Central Nervous System and Brain (IM, 29 minutes, VHS)
The Human Brain in Situ (FHS, 19 minutes, VHS, DVD)
The Human Nervous System: Brain and Cranial Nerves Videotape (BC, 28 minutes, VHS)
Neuroanatomy (UL, 19 minutes, VHS)
Sheep Brain Dissection (WNS, 22 minutes, VHS)

Advance Preparation

1. Make arrangements for appropriate storage, disposal, and cleanup of dissection materials. Check with the Department of Health or the Department of Environmental Protection, or their counterparts, for state regulations.
2. Designate a disposal container for organic debris, and a dishwashing area with hot soapy water, sponges, and a lab disinfectant such as Wavicide-01 (Carolina) for washing down the lab benches.
3. Set out disposable gloves and safety glasses.
4. Set out dissectible human brain models (ideally one per group), and preserved human brains.
5. Set out dissection kits, dissection trays, and sheep brains with meninges and cranial nerves intact.
6. Set up a videotape viewing area.

7. For testing cranial nerve function, set out dropper bottles of oil of cloves and vanilla, eye chart, ophthalmoscope, penlight, safety pins, mall probes (hot and cold), cotton, salty, sweet, sour, and bitter solutions, cotton swabs, ammonia, tuning forks, and tongue depressors. Set out autoclave bag for disposables.

Comments and Pitfalls

1. Students who are not careful readers confuse or do not distinguish between *cerebellar* and *cerebral.*
2. Hasty removal of the meninges removes the pituitary gland before its connection to the brain by the infundibulum can be established; occasionally even the optic chiasma is lost. Encourage the students to go slowly and use the scalpel sparingly.
3. The arachnoid meninx may be hard to identify, as it is usually poorly preserved.

Answers to Questions

Dissection: The Sheep Brain (pp. 120–122)

1. The sheep's cerebral hemispheres are smaller than those of the human.

Ventral Structures

1. The olfactory bulbs are larger in the sheep. The sense of smell is more important to the sheep than it is to humans for both protection and locating food.

Dorsal Structures

1. The cerebrum is not as deep.

3. The corpora quadrigemina are reflex centers for visual and auditory stimuli.

Spinal Cord and Spinal Nerves

exercise **15**

Time Allotment: $1^{1}/_{2}$ hours (1 hour can be completed outside of lab).

Multimedia Resources: See Appendix A for a list of multimedia resource distributors. See Exercise 5 for histology listings.

Brain and Nervous System: Your Information Superhighway (FHS, 31 minutes, VHS, DVD)

The Central Nervous System and Brain (IM, 29 minutes, VHS)

The Human Nervous System: Spinal Cord and Nerves Videotape (BC, 28 minutes, VHS)

The Peripheral Nervous System (UL, 29 minutes, VHS)

Advance Preparation

1. Make arrangements for appropriate storage, disposal, and cleanup of dissection materials. Check with the Department of Health or the Department of Environmental Protection, or their counterparts, for state regulations.
2. Designate a disposal container for organic debris, and a dishwashing area with hot soapy water and sponges. Provide a lab disinfectant such as Wavicide-01 (Carolina) for washing down the lab benches.
3. Set out disposable gloves and safety glasses.
4. Set out dissection tools, trays, petri dishes, and spinal cord sections from cow specimens or saved from the brain dissection.
5. Set out charts and models of the spinal cord, and colored pencils.
6. Set out slides of spinal cord cross section, lens paper, and lens cleaning solution.
7. Set out dissecting microscopes. Have compound microscopes available.

Comments and Pitfalls

1. Students may have trouble distinguishing between gray and white matter in the spinal cord dissection. A drop or two of methylene blue stain with a water rinse may help.

Answers to Questions

Dissection: Spinal Cord (p. 126)

1. The third meningeal layer is the pia mater, which adheres closely to the surface of the brain and spinal cord.
2. The posterior horns are more tapered than the anterior horns.
3. The central canal is more oval than circular. It is lined with ependymal cells. In a living specimen it contains cerebrospinal fluid. Students may observe that the posterior medial sulcus touches the posterior gray commissure (gray matter) of the spinal cord. Neuron cell bodies can be seen in the ventral horn of the spinal cord. The large neurons are motor neurons; others are association neurons. The dorsal root ganglions contain sensory neurons.

Human Reflex Physiology

exercise

16

Time Allotment: 1 hour.

Multimedia Resources: See Appendix A for a list of multimedia resource distributors.
Decision (FHS, 28 minutes, VHS, DVD)
Reflexes and Synaptic Transmission (UL, 29 minutes, VHS)

Solution:
Bleach Solution, 10%
Measure out 100 milliliters of bleach.
Add water to a final volume of 1 liter.

Advance Preparation

1. Fill a large laboratory bucket with *10% bleach solution* and set out an autoclave bag for disposable items. Set out wash bottles of *10% bleach solution.*
2. For each group, set out a reflex hammer, a cot (if available), a small piece of sterile absorbent cotton, a tongue depressor, a metric ruler, a 12-inch ruler, and a flashlight.

Comments and Pitfalls

1. Pupillary reflexes are more easily tested on subjects with light-colored irises.
2. Students do not always distinguish between the general term "pupillary reflexes" and the pupillary light reflex. Emphasize that the pupillary light reflex and the consensual response are both examples of pupillary reflexes.

Answers to Questions

Activity 1: Initiating Stretch Reflexes (pp. 134–135)

1. The leg swings forward as the quadriceps muscles contract. (The hamstrings are reciprocally inhibited.) The femoral nerve is carrying the impulses.
2. Fatigue results in a less vigorous response. Muscle function is responsible. This is probably due to changes in pH, ATP, and Ca^{2+} levels in the muscle. Excitation-contraction coupling is hindered, reducing the response of the muscle cells to nervous stimulation.

Muscle cells follow the all-or-none law—they contract the maximal amount for given physiological conditions, or not at all.

3. Plantar flexion due to the contraction of the triceps surae (gastrocnemius and soleus muscles) is the result. Contraction of the gastrocnemius muscle usually results in flexion of the knee.

Activity 2: Initiating the Plantar Reflex (p. 135)

The normal response is downward flexion (curling) and adduction of the toes. This is a normal plantar reflex.

Activity 3: Initiating the Corneal Reflex (p. 136)

The subject blinks. The function is to protect the eye. The subject experiences discomfort (if not pain) because the cornea lacks pressure receptors but is richly supplied with pain receptors.

Activity 4: Initiating the Gag Reflex (p. 136)

The posterior pharyngeal walls elevate as pharyngeal muscles contract, and the subject gags.

Activity 5: Initiating Pupillary Reflexes (p. 136)

4. The left pupil contracts (the pupillary light reflex).
5. The right pupil also contracts. The contralateral (consensual) reflex indicates that there is some connection between the pathways for each eye. This is a test of the parasympathetic nervous system. These responses protect the retina from damage by bright light.

exercise

The Special Senses

17

Time Allotment: 2–3 hours.

Multimedia Resources: See Appendix A for a list of multimedia resource distributors. See Exercise 5 for histology listings.

Balance (NIMCO, 28 minutes, VHS)
The Ear: Hearing and Balance (IM, 29 minutes, VHS)
The Eye: Structure, Function, and Control of Movement (FHS, 54 minutes, VHS, DVD)
The Eye: Vision and Perception (UL, 29 minutes, VHS)
Eyes and Ears (FHS, 28 minutes, VHS, DVD)
Hearing (FHS, 19 minutes, VHS, DVD)
Mystery of the Senses (CBS, 5-part series, 30 minutes each, VHS)
- *Hearing*
- *Smell*
- *Taste*
- *Vision*
- *Touch*

Now Hear This (NIMCO, 30 minutes, VHS)
Optics of the Human Eye Series (UL, 4-part series, VHS)
The Senses (FHS, 20 minutes, VHS, DVD)
The Senses of Smell and Taste (NIMCO, 28 minutes, VHS)
The Senses: Skin Deep (FHS, 26 minutes, VHS, DVD)
Sheep Eye Dissection (WNS, 15 minutes, VHS)
Taste (FHS, 23 minutes, VHS, DVD)
Taste and Smell (NIMCO, 30 minutes, VHS)
Visual Reality (NIMCO, 30 minutes, VHS)

Advance Preparation: The Eye and Vision

1. Make arrangements for appropriate storage and disposal of dissection materials. Check with the Department of Health or the Department of Environmental Protection for state regulations.
2. Designate a disposal container for organic debris and a dishwashing area with hot soapy water and sponges. Provide lab disinfectant such as Wavicide-01 (Carolina) for washing down the lab benches.

3. Set out disposable gloves and safety glasses.
4. Set out dissecting kits, dissecting pans, and preserved cow or sheep eyes. Plan for groups of two or individual dissections.
5. Set out dissectible eye models and eye anatomy charts.
6. Hang up a Snellen eye chart in a well-lit part of the room. Measure back 20 feet from the chart and mark the distance on the floor with masking tape.
7. Set out Ishihara's color-blindness plates (Carolina).

Comments and Pitfalls

1. Preserved cow eyes are often misshapen, and inexperienced students may need help locating and identifying the cornea at the beginning of the dissection.
2. For demonstration of the blind spot, emphasize that the dot disappears when the right eye is tested, and the X disappears when the left eye is tested. Some student is sure to claim that he/she has no blind spot in the left eye as the dot never disappeared!

Advance Preparation: The Ear, and Hearing and Balance

1. Set out dissectible ear models and ear anatomy charts.
2. Set up a compound microscope with a demonstration slide of the cochlea (Ward's).
3. For each group, set out tuning forks (Ward's), rubber mallet, absorbent cotton, a ticking pocket watch or small clock, and a 12-inch ruler.
4. Set out otoscopes, disposable otoscope tips, alcohol swabs, and an autoclave bag.

Comments and Pitfalls

1. It is often difficult to find an area quiet enough to get good results with the acuity and sound localization tests. An empty lab or a quiet corner of the hallway might be used.
2. Students should be reminded to simulate conductive deafness while performing the Weber test. Although it is not a specific assignment, they'll be asked for results in the Review Sheets.
3. Remind the students to strike the tuning forks with the rubber mallet and not against the lab bench.
4. Be sure the students understand how to evaluate the direction of nystagmus before the subject spins. Also remind the subject to keep his or her eyes open!

Advance Preparation: Smell and Taste

1. Set out for each group paper towels, a small mirror, a small dish of granulated sugar, absorbent cotton, four cotton-tipped swabs, one larger paper cup per person, dropper bottles of oil of cloves and oil of wintergreen and oil of peppermint (or corresponding flavors from the condiment section of the supermarket), a flask of distilled water, a paper plate, and chipped ice.
2. Set out a disposable autoclave bag, toothpicks, and disposable gloves.

3. Prepare a plate of cubed food items such as cheese, apple, raw potato, dried prunes, banana, raw carrot, and hard-cooked egg white. These foods should be in an opaque container; a foil-lined egg carton works well. Keep covered and refrigerated until used.

Comments and Pitfalls

1. Some students may have difficulty getting their noses to adapt to the aromatic oil. Be sure they are following directions carefully and are patient.
2. Subjects for the food tests should not be allowed to see the food.
3. Remind students to use toothpicks to select food cubes. Caution students to alert the instructor and group members about food allergies.

Answers to Questions

Activity 1: Identifying Accessory Eye Structures (p. 139)

Right eye: medial rectus

Left eye: lateral rectus (and on occasion the superior or inferior oblique)

Left and right eye: superior rectus muscle

Dissection: The Cow (Sheep) Eye (pp. 141–142)

6. The optic disc.

Activity 8: Demonstrating Reflex Activity of Intrinsic and Extrinsic Eye Muscles (p. 145)

Accommodation Pupillary Reflex

As the eye focuses on printed material, the pupil constricts. This reduces divergent light rays and aids in formation of a sharper image. It also restricts the amount of light entering the eye.

Convergence Reflex

The eyeballs will both move medially to focus on the object. This reflex keeps the image focused on the fovea.

Activity 12: Conducting Laboratory Tests of Hearing (p. 148)

Acuity Test

The threshold is indefinite.

Sound Localization

No, the sound is less easily located if the source is equidistant from both ears. Sound arriving from spots equidistant from both ears arrives at each ear at the same time and with equal loudness. This does not provide enough information to adequately locate the position of the source.

Activity 13: Conducting Laboratory Tests on Equilibrium (pp. 150–151)

Balance Test

1. Nystagmus is not present.

Romberg Test

2. Gross swaying movements are not usually observed when the eyes are open.
3. Side-to-side movement increases.
4. Front-to-back swaying occurs. The equilibrium apparatus and proprioceptors are probably functioning normally. Visual information is lacking and the result is increased swaying. Equilibrium and balance require input from a number of receptors including proprioceptors, the vestibular apparatus, and the eyes.

Activity 14: Identification of Papillae on the Tongue (p. 152)

It is easiest to identify fungiform and circumvallate papillae.

Activity 15: Stimulating Taste Buds (p. 153)

Substances must be in aqueous solution to stimulate the taste buds.

Activity 16: Examining the Combined Effects of Smell, Texture, and Temperature on Taste (pp. 153–154)

No, some foods can be identified fairly easily by texture. The sense of smell is most important when foods do not have an easily recognizable and unique texture. For example, it is hard to differentiate between raw apple and raw potato.

Effect of Olfactory Stimulation

2. It is hard to distinguish the flavor with the nostrils closed.
3. With the nostrils open it is easy to identify the oil.
6. The subject usually identifies the oil held at the nostrils.
7. Smell seems to be more important for identification in this experiment.

Effect of Temperature

The cold from the ice will inhibit the nerve endings on the tongue, therefore the subject won't be able to identify the food.

exercise

Functional Anatomy of the Endocrine Glands

18

Time Allotment: 1 hour.

Multimedia Resources: See Appendix A for a list of multimedia resource distributors. See Exercise 5 for histology listings.

Body Chemistry: Understanding Hormones (FHS, 3-part series, 50 minutes each, VHS, DVD)

Hormonally Yours
Hormone Heaven?
Hormone Hell

The Endocrine System (IM, UL, WNS, 17 minutes, VHS)
Glands and Hormones (NIMCO, 30 minutes, VHS)
Hormones: Messengers (FHS, 27 minutes, VHS, DVD)
The Hypothalamus and Pituitary Glands (UL, 29 minutes, VHS)
The Neuroendocrine System (IM, UL, 29 minutes, VHS)
The Pancreas (UL, 29 minutes, VHS)
Selected Actions of Hormones and Other Chemical Messengers Videotape (BC, 15 minutes, VHS)

InterActive Physiology® 8-System Suite–Endocrine System (BC, CD-ROM or www.interactivephysiology.com)
Orientation
Endocrine System Review
Biochemistry, Secretion, and Transport of Hormones
The Actions of Hormones on Target Cells
The Hypothalamic-Pituitary Axis
Response to Stress

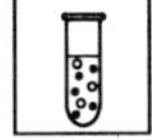

Solutions:

Glucose, 20%

Weigh out 200 grams of glucose. Add distilled water to a final volume of 1 liter.

Insulin, 400 IU/100 milliliter H_2O

Weigh out 16 milligrams of zinc-stabilized insulin (25 IU/milligram dry weight—ICN). Add water to a final volume of 100 milliliters.

Advance Preparation

1. Set out human torso models and anatomical charts of the human endocrine system.
2. Set out compound microscopes with slides of the thyroid gland and differentially stained pancreas tissue.
3. Prepare a solution of *20% glucose* and store it in the refrigerator. Purchase or order enough small fish to supply one per group. Set up an aquarium or large beaker with an air stone to maintain the fish.
4. For each group set out a 250-milliliter bottle of *20% glucose*, a dropper bottle of *insulin*, a glass marking pencil, and two finger bowls.

Comments and Pitfalls

1. The fish may become very agitated and jump out of the bowl or beaker.
2. This exercise may be upsetting to some students, even though the fish recover.

Answers to Questions

Activity 1: Examining the Microscopic Structure of the Thyroid Gland (pp. 156–157)

The different appearances are due to the fact that the gland has no duct and stores its secretion within the follicles. Thus, when the gland is inactive, it appears larger. In the active state, hormone is emptied into the blood stream, and it therefore appears smaller.

Activity 3: Examining the Microscopic Structure of the Pancreas to Identify Alpha and Beta Cells (p. 157)

2. Alpha cells produce glucagon and beta cells synthesize insulin.

Activity 4: Observing the Effects of Hyperinsulinism (pp. 157–158)

2. The fish often becomes very agitated, then loses its sense of balance just before becoming unconscious.
3. The fish will regain consciousness and right itself. The recovery time varies.

exercise

Blood

19

Note: For safety reasons, many instructors make the blood tests optional or try to provide alternative experiments. Substituting dog blood, as suggested below, is one option; using prepared slides or artificial blood are others.

Time Allotment: 2 hours.

Multimedia Resources: See Appendix A for a list of multimedia resource distributors. See Exercise 5 for histology listings.
Bleeding and Coagulation (FHS, 31 minutes, VHS, DVD)
Blood (UL 22 minutes, VHS)
Blood (FHS, 20 minutes, VHS, DVD)
Blood is Life (FHS, 25 minutes, VHS, DVD)

Blood and Immunity (CE, LP, CD-ROM)

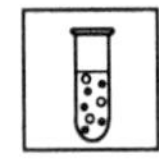

Solution:
Bleach Solution, 10%
Measure out 100 milliliters of household bleach. Add water to a final volume of 1 liter.

Advance Preparation

1. Set out safety glasses, lens paper, lens cleaning solution, and immersion oil. Have compound microscopes available. Set out any available models and charts of blood cells.
2. Set out prepared slides of neutrophils, eosinophils, basophils, lymphocytes, and monocytes.
3. Set up the following supply areas (if all tests are to be done). Ideally there should be at least one set of solutions for each lab bench and enough of the other supplies for each student to do each test. If equipment must be shared, it should be washed in hot soapy water and rinsed in *10% bleach solution* after each use.

   General supply area:

   a. For instructors using student blood samples, set out sterile blood lancets, alcohol wipes, and absorbent cotton balls. Set up a disposable autoclave bag for all disposable items, and a laboratory bucket or battery jar of *10% bleach solution* for glassware.

For each lab group, set out a 250-milliliter beaker of *10% bleach solution* (for used slides), spray bottles of *10% bleach solution,* clean microscope slides (two per member of the group), wide-range pH paper, test tube and test tube rack, nonhemolyzed plasma obtained fresh from an animal hospital or prepared by centrifuging animal (e.g., cattle or sheep) blood obtained from a biological supply house, timers, and disposable gloves.

b. For instructors using heparinized dog blood, set out heparinized dog blood, glass rods, and all the materials listed in paragraph a, except the sterile lancets and alcohol wipes.

Hematocrit supply area:

Set out heparinized capillary tubes, microhematocrit centrifuge and reading gauge or millimeter ruler, and capillary tube sealer (Carolina) or modeling clay.

Hemoglobin-determination supply area:

Set out hemoglobinometer and hemolysis applicator or Tallquist scales.

Coagulation time supply area:

Set out nonheparinized capillary tubes, fine triangular files, and paper towels.

Blood-typing supply area:

Set out blood-typing sera, Rh-typing boxes (if used), wax markers, toothpicks, and blood test cards or slides (Carolina). If you are using Ward's artificial blood, set out simulated blood and antibodies provided with the kit.

Comments and Pitfalls

1. If human blood samples are provided, disposable gloves and safety glasses should be worn at all times. If student samples are used, be sure students use only their own blood. Emphasize instructions for proper care or disposal of items used in the blood tests (see Human Anatomy & Physiology Laboratory Safety Guidelines, p. v of this *Instructor's Guide* and on the inside cover of the laboratory manual). Be sure that sharp objects such as lancets are discarded in such a way that they will not protrude through the autoclave bag. Petri dishes that can be taped shut or coffee cans with slits cut in the plastic tops can serve this purpose.
2. If student blood samples are used, have the students plan their work so that a minimum number of pricks are necessary. Obtaining enough blood is the usual problem. Be sure that students' hands are warm before trying to obtain blood, and that they follow the advice in the laboratory manual. Emphasize that the capillary tube should be held in a horizontal position with the tip in the drop of blood.
3. It is nearly impossible to prick your own finger to draw blood unless an automatic device such as the Medi-Let Kit (Carolina) is used. Students are often careless with the lancets since they are concentrating on obtaining blood for several different tests. Emphasize the importance of proper disposal. This is particularly important when using the Medi-Let, as it is difficult to distinguish a used lancet with a replaced cap from a new, unused one.

4. Obviously, heparinized blood samples may not be used for the coagulation time experiment.
5. Several problems may arise with the slides. Student-prepared blood smears tend to be too thick; be sure they understand the technique before starting. Warn against allowing the slide to dry with the stain on it.
6. A good color plate of the blood cells will help with identification. Also, point out the typical percentages of each cell type. Many students initially identify large numbers of cells as basophils.
7. Emphasize that the coagulation-time test must be started as soon as blood is drawn up into the capillary tube. Have students hold the tubes with paper towels when breaking them to avoid cuts.
8. Blood typing may be done here, but it may be easier to explain if it is done after the immune system has been discussed.

exercise

Anatomy of the Heart

20

Time Allotment: 1 1/2 hours.

Multimedia Resources: See Appendix A for a list of multimedia resource distributors. See Exercise 5 for histology listings.

The Circulatory System: Two Hearts that Beat as One (FHS, 28 minutes, VHS, DVD)
Human Cardiovascular System: The Heart Videotape (BC, 25 minutes, VHS)
Life Under Pressure (FHS, 26 minutes, VHS, DVD)
The Mammalian Heart (AIMS, 15 minutes, VHS, DVD)
Sheep Heart Dissection Video (WNS, 14 minutes, VHS)

InterActive Physiology® 8-System Suite–Cardiovascular System (BC, CD-ROM or www.interactivephysiology.com)
Anatomy Review: The Heart
Intrinsic Conduction System

Advance Preparation

1. Make arrangements for appropriate storage, disposal, and cleanup of dissection materials. Check with the Department of Health or the Department of Environmental Protection, or their counterparts, for state regulations.
2. Set out disposable gloves and safety glasses.
3. Set out dissecting kits, dissecting pans, glass probes, and preserved sheep hearts (one for each group).
4. Set out dissectible heart and cardiac muscle models, red and blue pencils, and heart anatomy charts.
5. Set out a compound microscope with prepared slides of cardiac muscle (longitudinal section).
6. Set out an X ray of the human thorax (if available) and an X ray viewing box.

Comments and Pitfalls

1. Some sheep hearts are sold with the pericardial sac removed. If possible, order sheep hearts with intact pericardial sacs (Ward's).

2. Students often confuse the base and apex of the heart.
3. Be sure students have correctly identified the left ventricle of the heart as a landmark before they begin the dissection. As with all dissections, urge students to be cautious with the scalpel.
4. Many preserved hearts have the venae cavae and pulmonary veins completely removed, leaving large holes in the walls of the atria. This will make it difficult for students to answer some of the questions in the lab text. Purchase a dissected pig heart that has all the major vessels intact (Carolina), or refer students to models if necessary.
5. Provide students with extra mall probes to mark vessels as they are identified.

Answers to Questions

Dissection: The Sheep Heart (pp. 175–178)

2. The pericardium is attached to the base of the heart.
3. The visceral pericardium is much thinner than the tough two-layered sero-fibrous parietal pericardium. The visceral pericardium adheres tightly to the heart, while the parietal pericardium forms the outer sac surrounding the pericardial cavity.
8. The lumen of the vena cava is larger. The aorta has thicker walls. The aorta is capable of stretching and elastic recoiling, which helps to maintain pressure in the vessels. This requires strength and resilience. The vena cava is a low-pressure vessel for blood return to the heart, and is not subjected to large pressure fluctuations.
9. The right atrioventricular valve has three flaps.
10. The pulmonary semilunar valve closes when fluid fills the collapsed cuplike valves, causing them to bulge out into the lumen. The atrioventricular valves are flaps that swing closed as pressure in the chamber increases. They are prevented from opening backwards into the atria by the chordae tendineae attached to the papillary muscles.
12. The left ventricular cavity is much narrower than the right ventricular cavity. Papillary muscles and chordae tendineae are present in both cavities. The left atrioventricular valve has two cusps. The sheep valves are very similar to their human counterparts.

exercise

Anatomy of Blood Vessels

21

Time Allotment: 1 1/2 hours.

Multimedia Resources: See Appendix A for a list of multimedia resource distributors. See Exercise 5 for histology listings.

Circulation: A River of Life (WNS, 30 minutes, VHS)
The Circulatory System (IM, 23 minutes, VHS)
The Circulatory System: The Plasma Pipeline (FHS, 25 minutes, VHS, DVD)
Human Biology (FHS, 58 minutes, VHS, DVD)
Human Cardiovascular System: Blood Vessels Videotape (BC, 25 minutes, VHS)
Life Under Pressure (FHS, 26 minutes, VHS, DVD)
Pumping Life—The Heart and Circulatory System (WNS, 20 minutes, VHS)

InterActive Physiology® 8-System Suite–Cardiovascular System (BC, CD-ROM or www.interactivephysiology.com)
Anatomy Review: Blood Vessel Structure and Function

Advance Preparation

1. Set out anatomical charts and/or models of human arteries and veins and the human circulatory system.
2. Set out anatomical charts of special circulations.
3. Set out compound microscopes with prepared slides of cross sections of arteries and veins.

Answers to Questions

Activity 4: Identifying Vessels of the Pulmonary Circulation (p. 188)

Labels for Figure 21.10 (clockwise, starting from the superior vena cava):

right pulmonary artery
pulmonary trunk
left pulmonary artery
left pulmonary veins
lobar arteris (left lung)
left ventricle
left atrium
right ventricle
right atrium
right pulmonary veins
lobar arteries (right lung)

Activity 5: Tracing the Pathway of Fetal Blood Flow (pp. 189–190)

Labels for Figure 21.11 (from top to bottom)

ductus arteriosus

foramen ovale

ductus venosus

umbilical vein

umbilical arteries

Activity 7: Tracing the Arterial Supply of the Brain (p. 191)

Labels for Figure 21.13 (from top to bottom):

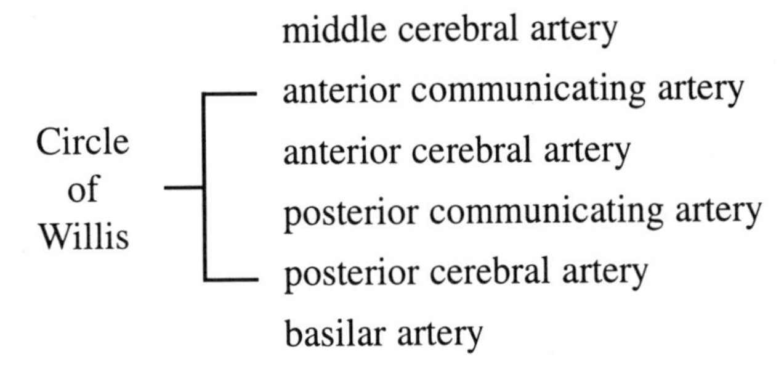

Human Cardiovascular Physiology—Blood Pressure and Pulse Determinations

exercise

22

Time Allotment: 2 hours (with some shared small-group data).

Multimedia Resources: See Appendix A for a list of multimedia resource distributors.
Human Biology (FHS, 58 minutes, VHS, DVD)
Life Under Pressure (FHS, 26 minutes, VHS, DVD)
The Physiology of Exercise (FHS, 15 minutes, VHS, DVD)

Blood and the Circulatory System NEO/LAB (LP, CD-ROM)

Interactive Physiology® 8-System Suite–Cardiovascular System (BC, CD-ROM or www.interactivephysiology.com)
Anatomy Review: Blood Vessel Structure and Function
Measuring Blood Pressure
Factors That Affect Blood Pressure
Blood Pressure Regulation
Autoregulation and Capillary Dynamic

Advance Preparation

1. Set out stethoscopes (both bell and diaphragm) and sphygmomanometers (two per group). Check the valves on the bulbs of the cuffs to be sure that air is released from the cuff when the valves are opened (replacement valves can be ordered).
2. Set out or ask students to bring watches with second hands. Also set out alcohol swabs and felt markers.
3. Set out one 0.4 m (16-inch) high bench (for women) and one 0.5 m (20-inch) high bench (for men). You may have to compromise with a .45 m (18-inch) bench. Set up a cot, if available.
4. Divide the class into small groups to collect data for Effect of Various Factors on Blood Pressure and Heart Rate. It may be hard to define *well-conditioned* and *poorly conditioned* subjects. A runner or a member of an athletic team might be compared to a more sedentary person (see Comments and Pitfalls, item 4).

Comments and Pitfalls

1. Most students in the health sciences will have no trouble with this lab, and in fact enjoy bringing their own stethoscopes and sphygmomanometers to lab if they are given advance notice.
2. If students have trouble hearing the heart sounds with the bell stethoscope, have them try the diaphragm model. This will be particularly helpful when trying to hear the split sounds. The sounds are louder with the bell stethoscopes, but placement must be more precise.
3. Caution students against overtightening the valve on the sphygmomanometer. If the air in the cuff can't be released, it is very painful to the subject. If the valve does stick, most cuffs can be undone even when filled with air. To avoid problems once the cuff is inflated, have students practice first with the bulb valve.
4. Students performing the Harvard step test should be carefully monitored to be sure that they step completely up and completely down at the prescribed rate. This can be very fatiguing. If the student population is fairly uniform it may be difficult to detect major differences between the *well-conditioned* and *poorly conditioned* individuals. Note: Another well-known test for fitness is the Schneider test (described in G. D. Tharp, *Experiments in Physiology*, 4th edition, Minneapolis, MN: Burgess Publishing Co., 1980). Try to compare people of the same general age and sex, and do not compare a smoker to a nonsmoker. Students who are aware that they have heart problems should be discouraged from acting as subjects.
5. Students who are testing the effects of venous congestion should be reminded to keep both arms quietly on the lab bench for the full 5 minutes. Check to be sure pressure is maintained at 40 mm Hg.

Answers to Questions

Activity 1: Auscultating Heart Sounds (p. 195)

3. The interval is about 0.5 second. It is about twice as long as the interval between the first and second heart sounds.

Activity 2: Palpating Superficial Pulse Points (pp. 195–196)

The carotid pulse point has the greatest amplitude, and the dorsalis pedis artery has the least. This is related to distance from the left ventricle of the heart.

Activity 5: Observing the Effect of Various Factors on Blood Pressure and Heart Rate (pp. 197–198)

6. Greater elevation of blood pressure is generally noted just after completion of exercise. Increased cardiac output during exercise results in increased systolic pressure. A poorly conditioned individual usually has a higher systolic pressure at the end of exercise, and it usually takes a longer time for the pressure to return to normal. A well-conditioned individual usually has a larger stroke volume and thus can pump more blood with fewer beats per minute than a poorly conditioned individual.

Activity 6: Examining the Effect of Local Chemical and Physical Factors on Skin Color (pp. 199–200)

Vasodilation and Flushing of the Skin Due to Lack of Oxygen

7. Stopping blood flow causes the hand to turn very pale. Weakness and a tingling sensation may be felt (variable). The skin flushes bright red immediately upon release of pressure and normal color is restored after several minutes or longer. There may be some lingering pain in the forearm region.

Effects of Venous Congestion

2. Slight pressure may be felt in the hand at the end of 5 minutes (variable). The veins are bulging and the hand has a mottled appearance, much darker in color than the control. Upon release of pressure, the veins deflate, and color and feeling return to normal.
3. Intensity of skin color (pink or blue) is related to the volume of blood in the area. The color is determined by the degree of oxygenation of the blood. In this experiment, venous blood gives a blue tint and arterial blood gives a pink tint.

Effect of Mechanical Stimulation of Blood Vessels of the Skin

Results will vary. A red streak develops with moderate pressure. With heavy pressure, a wider, darker, longer-lasting streak develops and may swell.

exercise

Anatomy of the Respiratory System

23

Time Allotment: 1 hour.

Multimedia Resources: See Appendix A for a list of multimedia resource distributors. See Exercise 5 for histology listings.

Breath of Life (FHS, 26 minutes, VHS, DVD)
The Dissection of the Thorax Series (UL, VHS)
  Part I. The Thoracic Wall (23 minutes)
  Part II. Pleurae and Lungs (24 minutes)
Human Respiratory System Videotape (BC, 25 minutes, VHS)
Lungs (Revised) (AIMS, 10 minutes, VHS)
The Respiratory System (UL, 26 minutes, VHS)
Respiratory System: Intake and Exhaust (FHS, 25 minutes, VHS, DVD)
Thorax (UL, 22 minutes, VHS)

InterActive Physiology® 8-System Suite–Respiratory System (BC, CD-ROM or www.interactivephysiology.com)
  Anatomy Review: Respiratory Structures

Advance Preparation

1. Set out human torso models, respiratory organ system model, larynx model, and/or charts of the respiratory system.
2. Set out a sheep pluck (fresh if possible), or set up an inflatable swine lungs kit (Nasco), disposable gloves, and a dissecting tray.
3. Arrange for a source of compressed air, or provide cardboard mouthpieces and a 2-foot length of rubber tubing. Set out an autoclave bag for disposal of the mouthpieces, if used.
4. Set out compound microscopes with prepared slides of the trachea and lung tissue.

Comments and Pitfalls

1. Many prepared slides of the trachea also include the esophagus. Remind the students that the trachea is held open by cartilaginous rings, while the esophagus is not. Showing a 35-millimeter slide or videodisc image of the section might be useful.

2. When using a preserved sheep pluck with a compressed air supply, be careful to avoid overinflation (leading to an explosion of preserved tissue)!
3. The inflatable swine lungs kit includes an inflation rack and tray, inflatable swine lungs, and a section of dried swine lung. The inflatable lungs will last for several years and give a much more dramatic response than that usually seen with the preserved lungs of the sheep pluck. An inflatable diseased lung is also available from Nasco and is excellent for comparison to a healthy lung.

exercise

Respiratory System Physiology

24

Time Allotment: 1 1/2 hours.

Multimedia Resources: See Appendix A for a list of multimedia resource distributors.
Breath of Life (FHS, 26 minutes, VHS, DVD)
Breathing (FHS, 20 minutes, VHS, DVD)
The Physiology of Exercise (FHS, 15 minutes, VHS, DVD)
Respiration (FHS, 15 minutes, VHS)
Respiratory System: Intake and Exhaust (FHS, 25 minutes, VHS, DVD)

InterActive Physiology® 8-System Suite–Respiratory System (BC, CD-ROM or www.interactivephysiology.com)
Anatomy Review: Respiratory Structures
Pulmonary Ventilation
Gas Exchange
Gas Transport
Control of Respiration

Advance Preparation

1. Set out the model lung.
2. For each group set out a tape measure, clear adhesive tape, nose clips, spirometer, disposable mouthpieces (enough for each member of the group), alcohol swabs, a paper bag, and a physiograph with a pneumograph and recording attachments. If a wet spirometer is used, be sure it is filled with distilled water according to the manufacturer's instructions. Set out a battery jar of 70% ethanol.
3. Set up a disposable autoclave bag.
4. Draw a class data chart on the blackboard to record TV (V_t), IRV, ERV, and VC.

Comments and Pitfalls

1. If a dry spirometer is used, the tidal volume readings are not very accurate. Somewhat better readings are obtained if the student exhales three times into the spirometer and divides the result by three.

2. Students will have to adjust the pneumograph until a good recording can be made. Be sure that it fits comfortably around the chest. Check all connections for a good fit, and if a tambour is used, be sure the rubber is intact. (This can be easily replaced using rubber sheeting. Use a good adhesive to reattach the clip.)
3. When using the pneumograph, be sure that students can correctly interpret the tracings. On some equipment, inspiration results in a downward deflection of the pointer. (Opposite the direction noted on the spirometer tracing in Figure 24.3).
4. Students may be confused about hyperventilation. The forced hyperventilation here results in a decreased breathing rate. Hyperventilation during psychological stress can produce a positive feedback situation, resulting in further hyperventilation. As hypocapnia increases, cerebral vessels constrict and increasingly acidotic conditions in the brain stimulate the medullary respiratory centers. Rebreathing breathed air in the latter case raises blood P_{CO_2}, reverses the cerebral vessel constriction, and stops the hyperventilation.

Answers to Questions

Activity 3: Visualizing Respiratory Variations (pp. 211–212)

4. During breath holding, the subject has the desire to expire. After a deep and forceful exhalation, the urge is to inspire. This may be explained by the Hering-Breuer reflex. Stretch receptors in the lungs are sensitive to extreme inflation and extreme deflation of the lungs. Impulses to the medulla oblongata initiate expiration or inspiration respectively.
5. The hyperventilation tracing should be similar in height and depth to the vital capacity tracing, but with an increased rate. After hyperventilation, the breathing rate slows down.
6. Breath-holding time increases after hyperventilation.
7. After 3 minutes of rebreathing breathed air, the ventilation rate increases. It is much faster than the breathing rate after hyperventilating.
9. Forced expiration results in dilation of the neck and face veins. Increased intrathoracic pressure reduces blood flow back to the heart, decreasing cardiac output. This results in increased cardiac rate (seen here as increased pulse rate).

Functional Anatomy of the Digestive System

exercise

25

Time Allotment: 2 hours.

Multimedia Resources: See Appendix A for a list of multimedia resource distributors. See Exercise 5 for histology listings.

Digestive System (WNS, 14 minutes, VHS)
Digestive System: Your Personal Power Plant (FHS, 20 minutes, VHS, DVD)
The Food Machine (NIMCO, 30 minutes, VHS)
The Guides to Dissection Series (UL, VHS)
Group V. The Abdomen (6 parts, 88.5 minutes total)
The Human Digestive System (AIMS, 18 minutes, VHS)
Human Digestive System Videotape (BC, 33 minutes, VHS)
Passage of Food Through the Digestive Tract (WNS, 8 minutes, VHS)

Solutions:

BAPNA, .01%
Weigh out .01 gram BAPNA. Add distilled water to a final volume of 100 milliliters.

Trypsin, 1%
Weigh out 1 gram trypsin. Add distilled water to a final volume of 100 milliliters.

Advance Preparation

1. Set out the dissectible torso model and anatomical charts of the human digestive system.
2. Set out models of a villus and the liver, if available; a jaw model; and/or a human skull.
3. Set out compound microscopes with prepared slides of the duodenum, mixed salivary glands, and liver.
4. Put a chart on the board for recording class results.
5. Set up the following supply areas:

   Enzyme action supply area (supply area 1):

   Set out test tubes, test tube racks, wax marking pencils, hot plates, graduated cylinders, 250-milliliter beakers, ice water bath, 37°C water bath, boiling chips (Carolina), hot plate, bile salts, Parafilm®, and dropper bottles of distilled water, 1% trypsin, 1% BAPNA solution, and vegetable oil.

Physical processes supply area (supply area 2):

Pitcher of water, paper cups, a stethoscope, disposable autoclave bag, and alcohol swabs.

6. Set up a VHS viewing area with the tape *Passage of Food Through the Digestive Tract* (Ward's) to allow independent viewing by students. This is an 8-minute film.

Comments and Pitfalls

1. This lab requires a great deal of organization and coordination on the part of the students. Emphasize the need for careful labeling and record keeping.
2. If a 37°C water bath is not available, incubate the tubes at room temperature and double the incubation time.
3. Enzyme activity can vary. Enzyme solutions should be prepared just before the lab and adjusted for appropriate activity.

Answers to Questions

Activity 1: Observing the Histological Structure of the Alimentary Canal Wall (p. 216)

Simple columnar epithelium

Activity 11: Demonstrating the Action of Bile on Fats (p. 225)

Emulsification occurs in the tubes containing bile salts.

Activity 12: Observing Movements and Sounds of Digestion (pp. 225–226)

3. Movement of the larynx ensures that its passageway is covered by the epiglottis.

Functional Anatomy of the Urinary System

exercise

26

Time Allotment: 1–2 hours.

Multimedia Resources: See Appendix A for a list of multimedia resource distributors. See Exercise 5 for histology listings.

Human Urinary System Videotape (BC, 23 minutes, VHS)
Kidney Functions (AIMS, 5 minutes, VHS, DVD)
The Kidney (FHS, 14 minutes, VHS)
The Urinary Tract: Water! (FHS, 28 minutes, VHS, DVD)

InterActive Physiology® 8-System Suite–Urinary System (BC, CD-ROM or www.interactivephysiology.com)
- Anatomy Review
- Glomerular Filtration
- Early Filtrate Processing
- Late Filtrate Processing

InterActive Physiology® 8-System Suite–Fluids and Electrolytes (BC, CD-ROM or www.interactivephysiology.com)
- Introduction to Body Fluids
- Water Homeostasis
- Electrolyte Homeostasis
- Acid/Base Homeostasis

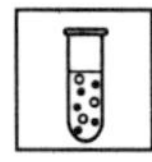

Solutions:

*Urine, Artificial Normal Human**
- 36.4 grams urea
- 15 grams sodium chloride
- 9.0 grams potassium chloride
- 9.6 grams sodium phosphate
- 4.0 grams creatinine
- 100 milligrams albumin

Add urea to 1.5 liters of distilled water. Mix until crystals dissolve. Add sodium chloride, potassium chloride, and sodium phosphate. Mix until solution is clear. The pH should be within the 5 to 7 pH range for normal human urine. Adjust pH, if necessary, with 1 *N* HCl or 1 *N* NaOH. Place a urine hydrometer in the solution and dilute with water to a specific gravity within the range of 1.015 to 1.025. This stock solution may be refrigerated for several weeks or frozen for months. Before use, warm to room temperature and add 4.0 grams creatinine and 100 milligrams of albumin for each 2 liters of solution.

*Urine, Glycosuria**
For a minimally detectable level of glucose, add a minimum of 600 milligrams of glucose to 1 liter of "normal" urine solution. For moderate to high glycosuria, add 2.5 to 5.0 grams of glucose to each liter of solution.

*Urine, Hematuria**
Add 1 milliliter of heparinized or defibrinated sheep blood to 1 liter of "normal" urine solution.

*Urine, Hemoglobinuria**
Add 2 milligrams of bovine hemoglobin to 1 liter of "normal" urine solution.

*Urine, Hyposthenuria**
Add distilled water to a sample of "normal" urine until the specific gravity approaches 1.005.

*Urine, Ketonuria**
Add a minimum of 100 milligrams of acetoacetic acid or at least 1 milliliter of acetone to 1 liter of "normal" urine solution.

*Urine, pH Imbalance**
Adjust "normal" urine to a pH of 4.0 to 4.5 with 1 *N* HCl for acid urine. Adjust "normal" urine to a pH of 8 to 9 with 1 *N* NaOH for alkaline urine.

*Urine, Proteinuria**
Add 300 milligrams or more of albumin per liter of "normal" urine solution. For severe renal damage, add 1 gram of albumin to each liter of solution.

*Urine, Whole Spectrum Pathological Artificial Human**
Mix appropriate amounts of abnormal condition reagents to 1 liter of "normal" urine solution.
Diabetes mellitus: glycosuria and ketonuria
Glomerular damage: proteinuria, hemoglobinuria, and hematuria

* From B. R. Shmaefsky, "Artificial Urine for Laboratory Testing," *American Biology Teacher* 52 (3), March 1990, pp. 170–172 (Reston, VA: National Association of Biology Teachers). Reprinted with permission.

Advance Preparation

1. Make arrangements for appropriate storage, disposal, and cleanup of dissection materials. Check with the Department of Health, the Department of Environmental Protection, or their counterparts for state regulations.

2. Set out disposable gloves and safety glasses.
3. Set out dissecting kits, dissecting pans, and pig or sheep kidneys.
4. Set out compound microscopes with slides of longitudinal sections of the kidney.
5. Set out the dissectible human torso and/or any anatomical charts and models of the urinary system, kidney, and nephron.
6. Set out sterile containers for collection of student urine samples or prepare *"normal" artifical urine* (about 1 liter for a class of 30 students). Prepare *"pathological" artificial urine* samples and number them.
7. Set out two laboratory buckets containing *10% bleach solution*, and a disposable autoclave bag. Put a flask of *10% bleach solution* and a sponge at each lab bench.
8. For each student in the class set out disposable gloves, five test tubes, a glass stirring rod, wide-range pH paper, a test tube rack, a medicine dropper, a urinometer cylinder and float, and combination strips (Chemstrip—Carolina or Multistix—Fisher).

Comments and Pitfalls

1. When preparing pathological samples, do not substitute sucrose for glucose. Vitamin C contamination will give false positive glucose tests. The artificial urine is suitable for test strips, but not for use with clinical analyzers (Shmaefsky, 1990).

exercise

Anatomy of the Reproductive System

27

Time Allotment: 1 hour.

Multimedia Resources: See Appendix A for a list of multimedia resource distributors. See Exercise 5 for histology listings.

The Guides to Dissection Series (UL, VHS)
Group VI. The Pelvis and Perineum (4 parts, 64 minutes total)
Human Biology (FHS, 58 minutes, VHS, DVD)
The Human Female Reproductive System (UL, 29 minutes, VHS)
The Human Male Reproductive System (UL, 29 minutes, VHS)
Human Reproductive Biology (FHS, 35 minutes, VHS, DVD)
Human Reproductive System Videotape (BC, 32 minutes, VHS)
Reproduction: Shares in the Future (FHS, 26 minutes, VHS, DVD)

Advance Preparation

1. Set out compound microscopes with slides of human sperm and of an ovary.
2. Set out models and anatomical charts of the male and female reproductive systems.

NAME ______________________ LAB TIME/DATE ______________

REVIEW SHEET
exercise
1

The Language of Anatomy

Surface Anatomy

1. Match each of the following descriptions with a key term, and record the term in front of the description.

Key: brachial, buccal, carpal, cervical, deltoid, digital, patellar, scapular

buccal 1. cheek

digital 2. referring to the fingers

scapular 3. shoulder blade region

carpal 4. wrist area

patellar 5. anterior aspect of knee

brachial 6. referring to the arm

deltoid 7. curve of shoulder

cervical 8. referring to the neck

Body Orientation, Direction, Planes, and Sections

2. Several incomplete statements are listed below. Correctly complete each statement by choosing the appropriate anatomical term from the key. Record the key terms on the correspondingly numbered blanks below.

Key: anterior, distal, frontal, inferior, lateral, medial, posterior, proximal, sagittal, superior, transverse

In the anatomical position, the umbilicus and knees are on the __1__ body surface; the calves and shoulder blades are on the __2__ body surface; and the soles of the feet are the most __3__ part of the body. The ears are __4__ and __4__ to the shoulders and __5__ to the nose. The breastbone is __6__ to the vertebral column (spine) and __7__ to the shoulders. The elbow is __8__ to the shoulder but __9__ to the fingers. The thoracic cavity is __10__ to the abdominopelvic cavity and __11__ to the spinal cavity. In humans, the ventral surface can also be called the __12__ surface; however, in quadruped animals, the ventral surface is the __13__ surface.

If an incision cuts the brain into superior and inferior parts, the section is a __14__ section; but if the brain is cut so that anterior and posterior portions result, the section is a __15__ section. You are told to cut a dissection animal along two planes so that the lungs are observable in both sections. The two sections that meet this requirement are the __16__ and __17__ sections.

1. *anterior*
2. *posterior*
3. *inferior*
4. *superior*
   medial
5. *lateral*
6. *anterior*
7. *medial*
8. *distal*
9. *proximal*
10. *superior*
11. *anterior*
12. *anterior*
13. *inferior*
14. *transverse*
15. *frontal*
16. *transverse*
17. *frontal*

3. A nurse informs you that she is about to give you a shot in the lateral femoral region. What portion of your body should you uncover?

Side of upper thigh

4. Correctly identify each of the body planes by inserting the appropriate term for each on the answer line below the drawing.

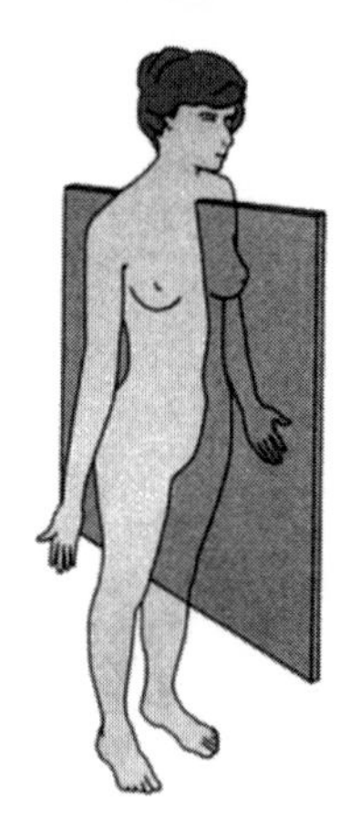

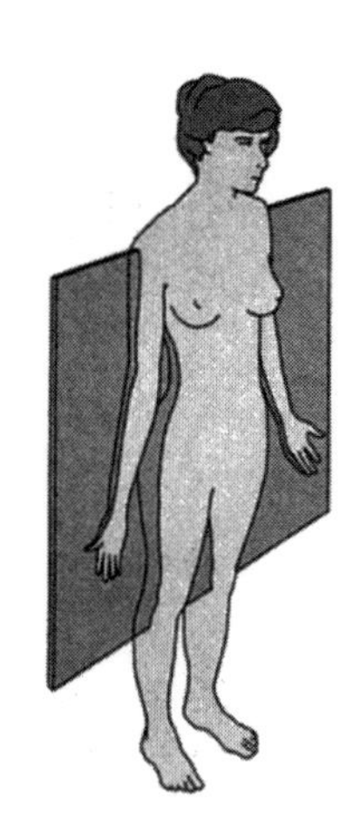

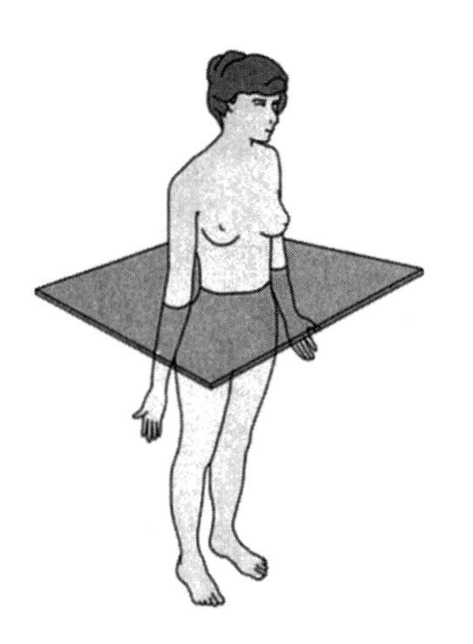

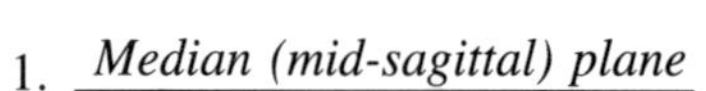

1. *Median (mid-sagittal) plane*
2. *Frontal*
3. *Transverse*

Body Cavities

5. Which body cavity would have to be opened for the following types of surgery? (Insert the key term(s) in the same-numbered blank. More than one choice may apply.)

Key: abdominopelvic, cranial, dorsal, spinal, thoracic, ventral

1. surgery to remove a cancerous lung lobe — 1. *thoracic/ventral*
2. removal of an ovary — 2. *abdominopelvic/ventral*
3. surgery to remove a ruptured disk — 3. *spinal/dorsal*
4. appendectomy — 4. *abdominopelvic/ventral*
5. removal of the gallbladder — 5. *abdominopelvic/ventral*

6. Correctly identify each of the described areas of the abdominal surface by inserting the appropriate term in the answer blank preceding the description.

hypochondriac region 1. overlies the lateral aspects of the lower ribs

umbilical region 2. surrounds the "belly button"

hypogastric region 3. encompasses the pubic area

epigastric region 4. medial region overlying the stomach

7. What are the bony landmarks of the abdominopelvic cavity?

Rib cage and pelvis

8. Which body cavity affords the least protection to its internal structures? *Abdominopelvic cavity*

NAME ______________ LAB TIME/DATE ______________

REVIEW SHEET
exercise
2

Organ Systems Overview

1. Using the key choices, indicate the body systems that match the following descriptions. Then, circle the organ systems (in the key) which are present in all subdivisions of the ventral body cavity.

Key: (cardiovascular) digestive endocrine; integumentary (lymphatic) muscular; (nervous) reproductive respiratory; skeletal urinary

urinary 1. rids the body of nitrogen-containing wastes

endocrine 2. is affected by removal of the adrenal gland

skeletal 3. protects and supports body organs; provides a framework for muscular action

cardiovascular 4. includes arteries and veins

endocrine 5. composed of "ductless glands" that secrete hormones

integumentary 6. external body covering

lymphatic 7. houses cells involved in body immunity

digestive 8. breaks down ingested food into its absorbable units

respiratory 9. loads oxygen into the blood

cardiovascular/endocrine 10. uses blood as a transport vehicle

muscular 11. generates body heat and provides for locomotion of the body as a whole

urinary 12. regulates water and acid-base balance of the blood

reproductive and *endocrine* 13. necessary for childbearing

integumentary 14. is damaged when you fall and scrape your knee

2. Using the above key, choose the *organ system* to which each of the following sets of organs or body structures belongs:

lymphatic 1. lymph nodes, spleen, lymphatic vessels

skeletal 2. bones, cartilages, ligaments

endocrine 3. thyroid, thymus, pituitary

respiratory 4. trachea, bronchi, alveoli

reproductive 5. uterus, ovaries, vagina

cardiovascular 6. arteries, veins, heart

3. Using the key below, place the following organs in their proper body cavity:

Key: abdominopelvic cranial spinal thoracic

abdominopelvic 1. stomach

thoracic 2. esophagus

abdominopelvic 3. large intestine

abdominopelvic 4. liver

spinal 5. spinal cord

abdominopelvic 6. urinary bladder

thoracic 7. heart

thoracic 8. trachea

cranial 9. brain

abdominopelvic 10. rectum

4. Using the organs listed in item 3 above, record, by number, which would be found in the following abdominal regions:

3, 6, 10 1. hypogastric region

3 2. right lumbar region

3 3. umbilical region

1, 4 4. epigastric region

3 5. left iliac region

1 6. left hypochondriac region

5. The five levels of organization of a living body, beginning with the cell, are: cell, *tissue*, *organ*, *organ system*, and organism.

6. Define *organ:* *A structure composed of two or more tissues that performs a specialized function*

7. Using the terms provided, correctly identify all of the body organs provided with leader lines in the drawings below. Then name the organ systems by entering the name of each on the answer blank below each drawing.

Key: blood vessels, brain, heart, kidney, nerves, sensory organ, spinal cord, ureter, urethra, urinary bladder

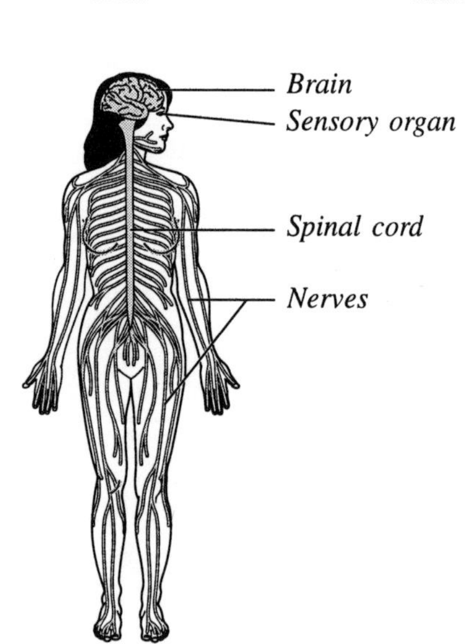

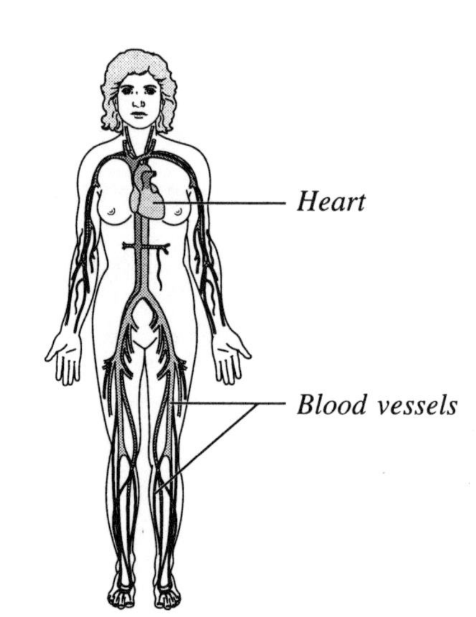

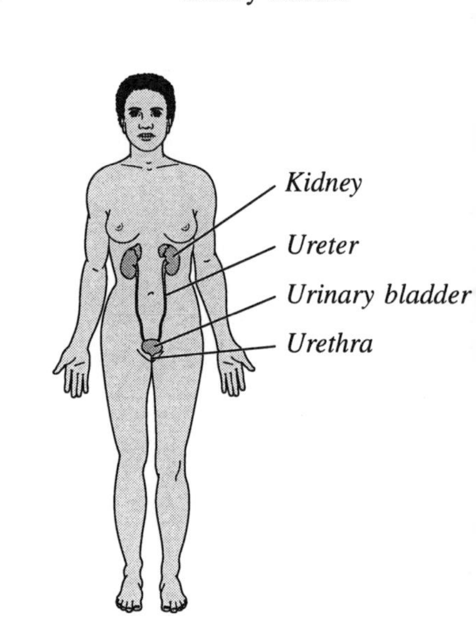

1. *Nervous* 2. *Cardiovascular* 3. *Urinary*

NAME ______________________ LAB TIME/DATE ______________________

REVIEW SHEET
exercise
3

The Cell—Anatomy and Division

Anatomy of the Composite Cell

1. Define the following:

Organelle: *"small organs"—The metabolic machinery of the cell organized to carry out specific activities for the cell as a whole*

Cell: *The structural and functional unit of all living things*

2. Identify the following cell parts:

plasma membrane 1. external boundary of cell; regulates flow of materials into and out of the cell

lysosomes 2. contains digestive enzymes of many varieties; "suicide sac" of the cell

mitochondria 3. scattered throughout the cell; major site of ATP synthesis

microvilli 4. slender extensions of the plasma membrane that increase its surface area

inclusions 5. stored glycogen granules, crystals, pigments, and so on

Golgi apparatus 6. membranous system consisting of flattened sacs and vesicles; packages proteins for export

nucleus 7. control center of the cell; necessary for cell division and cell life

centrioles 8. two rod-shaped bodies near the nucleus; the basis of cilia

nucleolus 9. dense, darkly staining nuclear body; packaging site for ribosomes

microfilaments 10. contractile elements of the cytoskeleton

endoplasmic reticulum 11. membranous system that has "rough" and "smooth" varieties

ribosomes 12. attached to membrane systems or scattered in the cytoplasm; synthesize proteins

chromatin 13. threadlike structures in the nucleus; contain genetic material (DNA)

peroxisome 14. site of detoxification of harmful chemicals

3. In the following diagram, label all parts provided with a leader line.

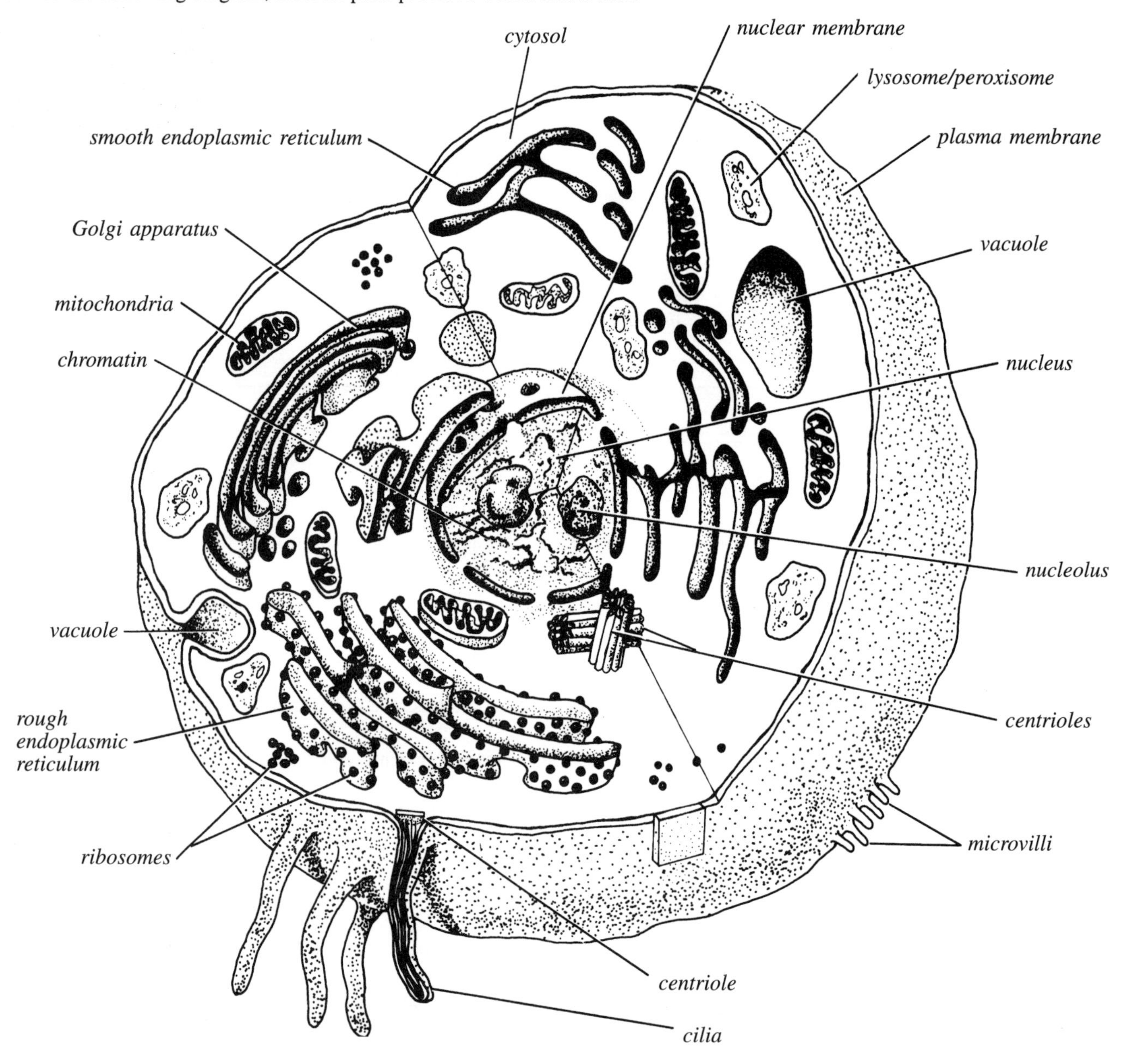

Differences and Similarities in Cell Structure

4. For each of the following cell types, on line (a) list *one* important *structural* characteristic observed in the laboratory. On line (b) write the *function* that the structure complements or ensures.

squamous epithelium a. *thin layer, flattened cells; fit closely together*

b. *allows for easy diffusion and filtration*

sperm a. *compact and streamlined; flagella*

b. *able to propel itself in a short time*

smooth muscle a. *non-striated, elongated, spindle-shaped*

b. *provides a long axis for contraction*

red blood cells a. *biconcave disc; lacks nucleus; have few organelles*

b. *easy gas exchange; more room for hemoglobin to transport oxygen*

5. What is the significance of the red blood cell being anucleate (without a nucleus)? *Increase area to carry oxygen and for gas exchange; does not use oxygen (no metabolism)*

Did it ever have a nucleus? *Yes* When? *In the bone marrow while maturing*

Cell Division: Mitosis and Cytokinesis

6. Identify the three phases of mitosis shown in the following photomicrographs, and select the events from the key choices that correctly identify each phase. Write the key letters on the appropriate answer line.

Key:

a. Chromatin coils and condenses, forming chromosomes.
b. The chromosomes (chromatids) are V-shaped.
c. The nuclear membrane re-forms.
d. Chromosomes stop moving toward the poles.
e. Chromosomes line up in the center of the cell.
f. The nuclear membrane fragments.
g. The spindle forms.
h. DNA synthesis occurs.
i. Chromosomes first appear to be double.
j. Chromosomes attach to the spindle fibers.
k. The nuclear membrane(s) is absent.

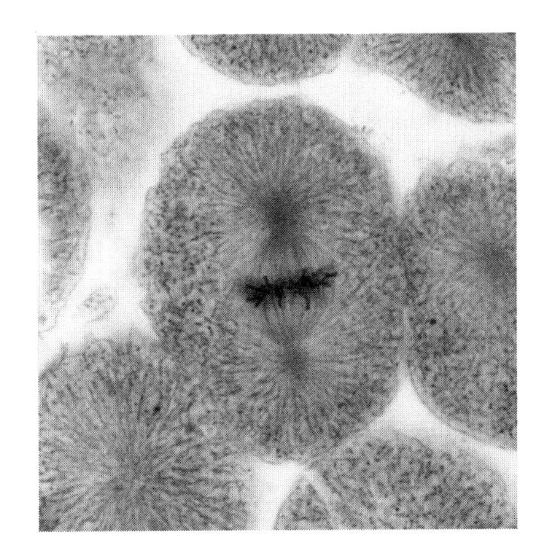

1. Phase: *Metaphase*

Events: *e*

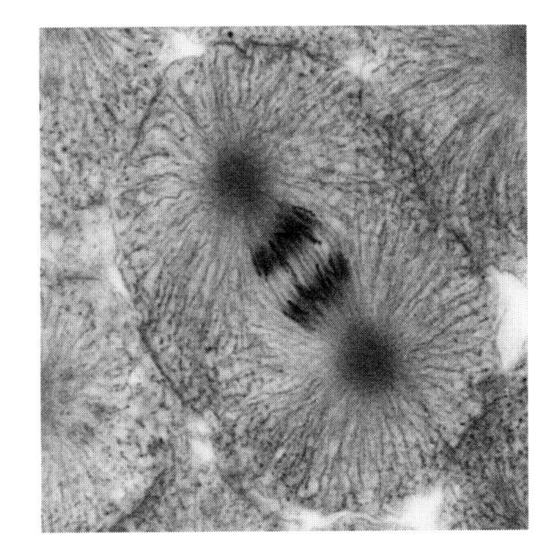

2. Phase: *Anaphase*

Events: *b d*

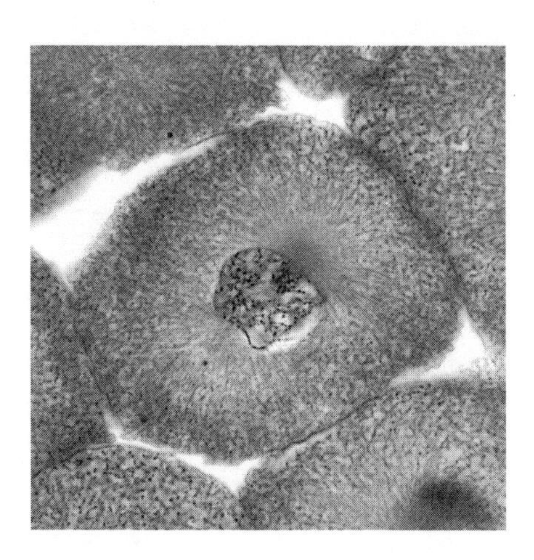

3. Phase: *Prophase*

Events: *a f g i j k*

7. What is the importance of mitotic cell division? *The function of mitotic cell division in the body is to increase the number of cells for growth and repair.*

NAME ____________________ LAB TIME/DATE ____________________

REVIEW SHEET
exercise
4

Cell Membrane Transport Mechanisms

Choose all answers that apply to items 1 and 2, and place their letters on the response blanks.

1. The motion of molecules *a d*.

a. reflects the kinetic energy of molecules
b. reflects the potential energy of molecules
c. is ordered and predictable
d. is random and erratic

2. Kinetic energy *b c e*.

a. is higher in larger molecules
b. is lower in larger molecules
c. increases with increasing temperature
d. decreases with increasing temperature
e. is reflected in the speed of molecular movement

3. What is the relationship between molecular size and diffusion rate? *Light molecules move more quickly than heavy ones*

What is the relationship between the temperature and diffusion rate? *Direct relationship (increase temperature = increase diffusion rate)*

4. The following refer to Activity 2 using dialysis sacs to study diffusion through nonliving membranes:

Sac 1: 40% glucose suspended in distilled water

Did glucose pass out of the sac? *yes* Test used to determine presence of glucose: *Benedict's solution*

Did the sac weight change? *yes* If so, explain the reason for its weight change: *Water diffused into sac*

Sac 2: 40% glucose suspended in 40% glucose

Was there net movement of glucose in either direction? *No*

Explanation: *Equal concentration of glucose in and out of sac*

Did the sac weight change? *No* Explanation: *No net movement of water or glucose*

Sac 3: 10% NaCl in distilled water

Was there net movement of NaCl out of the sac? *Yes*

Test used to determine the presence of NaCl: *Silver nitrate*

Direction of net osmosis: *Into sac*

Sac 4: Sucrose and Congo red dye in distilled water

Was there net movement of dye out of the sac? *No*

Was there net movement of sucrose out of the sac: *No*

Explanation: *Upon boiling, some of the sucrose bonds are hydrolyzed releasing glucose and fructose.*

Test used to determine the movement of sucrose into the beaker? *Using Benedict's test indicates the presence of glucose if sucrose passed through the membrane.*

Direction of net osmosis: *Into sac*

5. What single characteristic of the semipermeable membranes used in the laboratory determines the substances that can pass through them? *Size of molecules*

In addition to this characteristic, what other factors influence the passage of substances through living membranes?

Concentration gradient, molecules dissolving in a lipid bilayer, kinetic energy

6. A semipermeable sac containing 4% NaCl, 9% glucose, and 10% albumin is suspended in a solution with the following composition: 10% NaCl, 10% glucose, and 40% albumin. Assume that the sac is permeable to all substances except albumin. State whether each of the following will (a) move into the sac, (b) move out of the sac, or (c) not move.

glucose *a* water *b* albumin *c* NaCl *a*

7. The diagrams below represent three microscope fields containing red blood cells. Arrows show the direction of net osmosis. Which field contains a hypertonic solution? (If necessary, check your textbook for the meaning of *hypertonic, isotonic,* and *hypotonic* before tackling this one.) *c* The cells in this field are said to be *crenated*.

Which field contains an isotonic bathing solution? *b* Which field contains a hypotonic solution? *a*

What is happening to the cells in this field? *Lysing*

(a) (b) (c)

8. What is the driving force for filtration? *Hydrostatic pressure*

How does knowing this help you to explain why the filtration process examined in lab slowed down with time? *Less pressure*

9. Define *diffusion:* *The movement of molecules from a region of higher concentration to a region of lower concentration*

NAME ____________________ LAB TIME/DATE ____________________

REVIEW SHEET
exercise
5

Classification of Tissues

Tissue Structure and Function: General Review

1. Define *tissue:* *groups of cells that are similar in structure and function*

2. Use the key choices to identify the major tissue types described below.

Key: connective epithelium muscular nervous

epithelium 1. lines body cavities and covers the body's external surface

muscle 2. pumps blood, flushes urine out of the body, allows one to swing a bat

nervous 3. transmits waves of excitation

connective 4. anchors and packages body organs

epithelium 5. cells may absorb, protect, or form a filtering membrane

nervous 6. most involved in regulating body functions quickly

muscle 7. major function is to contract

connective 8. the most durable tissue type

connective 9. abundant nonliving extracellular matrix

nervous 10. forms nerves

Epithelial Tissue

3. On what bases are epithelial tissues classified? *Arrangement (relative number of layers); cell shape (squamous, cuboidal, columnar)*

4. How is the function of an epithelium reflected in its arrangement? *Function of epithelial cells is to form linings or covering membranes—reflected in the arrangement of fitting closely together to form intact sheets of cells*

5. Where is ciliated epithelium found? *Lines trachea, most of the upper respiratary tract*

What role does it play? *Motile cell projections that help to move substances along the cell surface*

6. Transitional epithelium is actually stratified squamous epithelium, but there is something special about it.

How does it differ structurally from other stratified squamous epithelia? *It differs from other stratified squamous epi's by being rounded*

How does this reflect its function in the body? *Has the ability to slide over one another to allow the organ to be stretched*

7. Respond to the following with the key choices.

Key: pseudostratified ciliated columnar, simple columnar, simple cuboidal, simple squamous, stratified squamous, transitional

stratified squamous epithelium 1. best suited for areas subject to friction

pseudostratified ciliated columnar 2. propels substances across its surface

simple squamous 3. most suited for rapid diffusion

simple cuboidal 4. tubules of the kidney

pseudostratified ciliated columnar 5. lines much of the respiratory tract

transitional 6. stretches

simple columnar epithelium 7. lines the small and large intestines

Connective Tissue

8. What is the makeup of the matrix in connective tissues? *Nonliving material composed of ground substance—(glycoproteins and large polysaccharide molecules) and fibers*

9. How are the functions of connective tissue reflected in its structure? *The function of connective tissue is protection, support and bind other tissues together—reflected in matrix for strength and fibers for support.*

10. Using the key, choose the best response to identify the connective tissues described below.

Key:
- adipose connective tissue
- areolar connective tissue
- dense fibrous connective tissue
- reticular connective tissue
- hyaline cartilage
- osseous tissue

dense fibrous connective ______ 1. attaches bones to bones and muscles to bones

osseous tissue ______ 2. forms your hip bone

areolar connective ______ 3. composes basement membranes; a soft packaging tissue with a jellylike matrix

hyaline cartilage ______ 4. forms the larynx and the costal cartilages of the ribs

hyaline cartilage ______ 5. firm matrix heavily invaded with fibers; appears glassy and smooth

osseous tissue ______ 6. matrix hard; provides levers for muscles to act on

adipose connective ______ 7. insulates against heat loss; provides reserve fuel

Muscle Tissue

11. The terms and phrases in the key relate to the muscle tissues. For each of the three muscle tissues, select the terms or phrases that characterize it, and write the corresponding letter of each term on the answer line.

Key:
- a. striated
- b. branching cells
- c. spindle-shaped cells
- d. cylindrical cells
- e. active during birth
- f. voluntary
- g. involuntary
- h. one nucleus
- i. many nuclei
- j. forms heart walls
- k. attached to bones
- l. intercalated discs
- m. in wall of bladder and stomach
- n. moves limbs, produces smiles
- o. arranged in sheets

Skeletal muscle: *f d i a n k*

Cardiac muscle: *j a b h l g (e) (d)*

Smooth muscle: *e m h c (g) o*

Nervous Tissue

12. In what ways are nerve cells similar to other cells? *Nucleus-containing cell with cytoplasm*

How are they different? *Their cytoplasm is drawn out into long extensions (cell processes)*

How does the special structure of a neuron relate to its function? *Allows a single neuron to conduct an impulse over relatively long distances*

For Review

13. Write the name of each tissue type in illustrations a through l, and label all major structures.

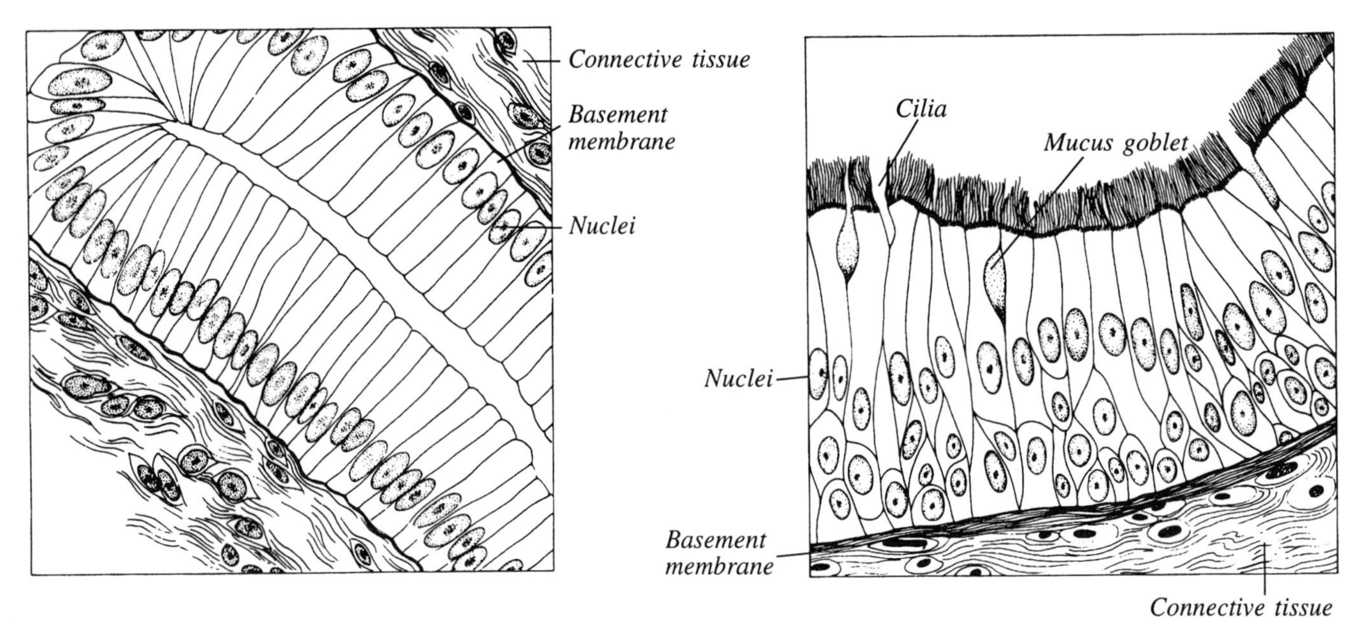

a. *Simple columnar epithelium*

b. *Pseudostratified ciliated columnar epithelium*

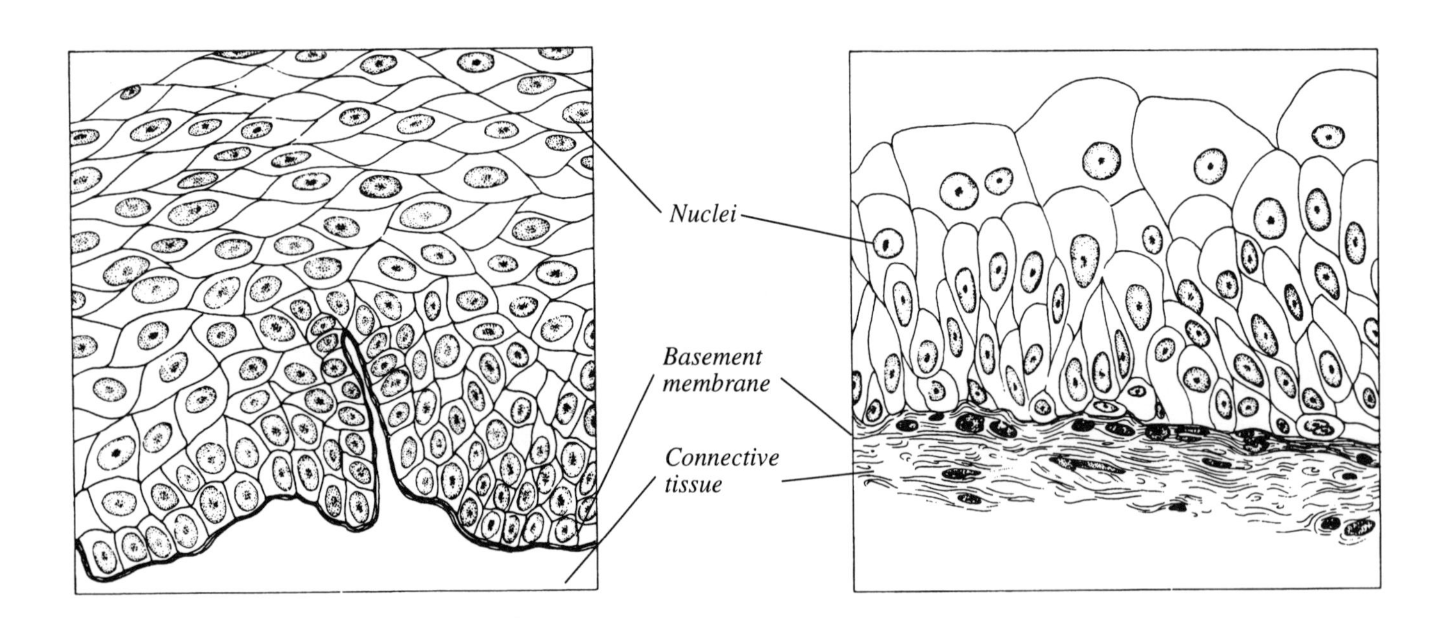

c. *Stratified squamous epithelium*

d. *Transitional epithelium*

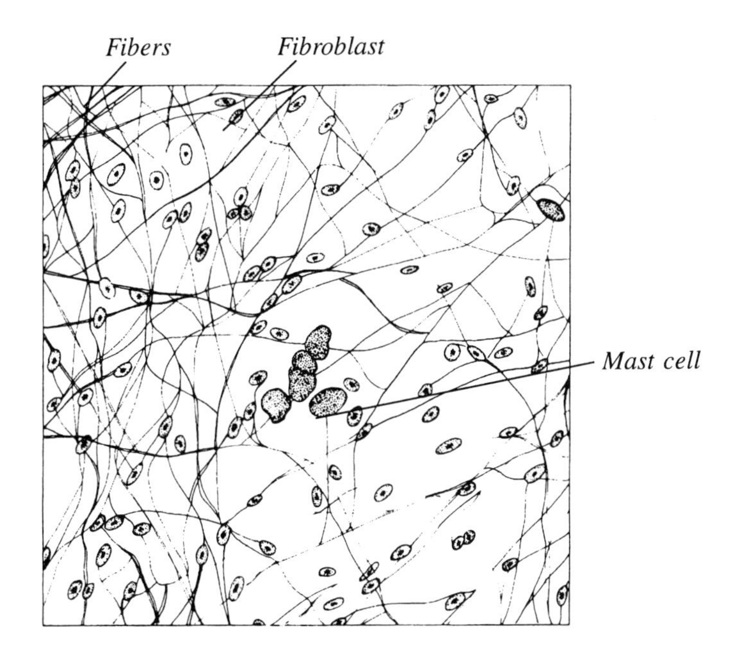

e. Areolar connective tissue

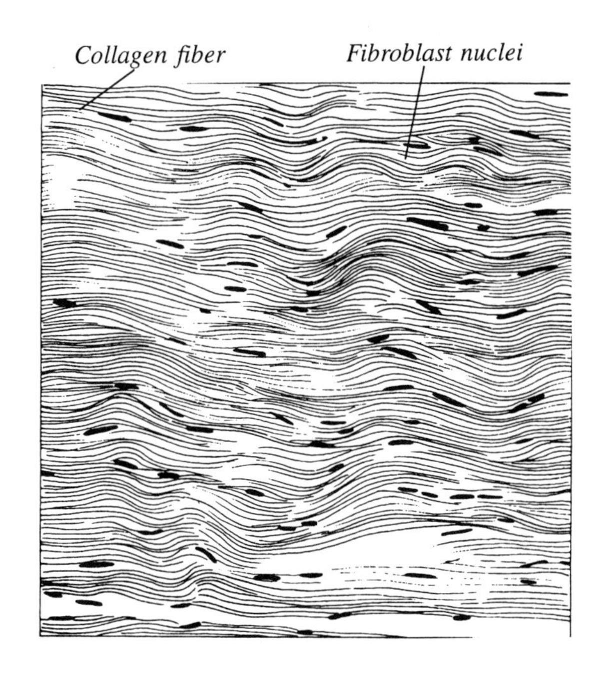

f. Dense connective tissue

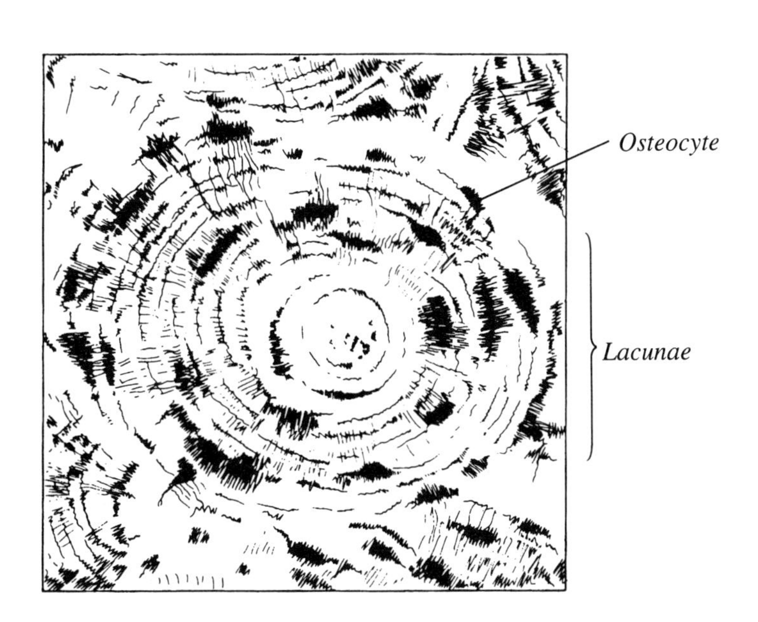

g. Osseous tissue

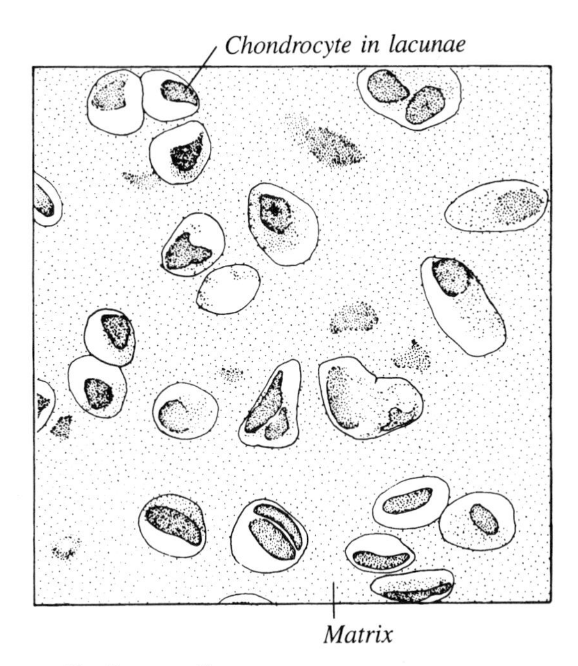

h. Hyaline cartilage

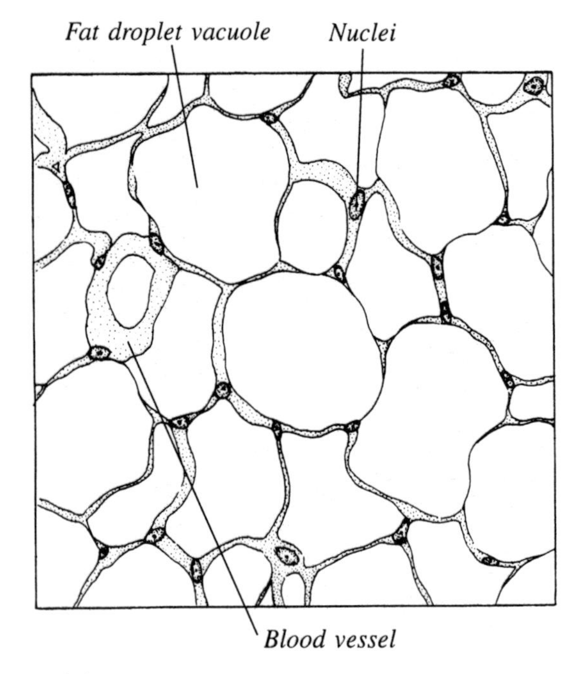

i. *Adipose tissue*

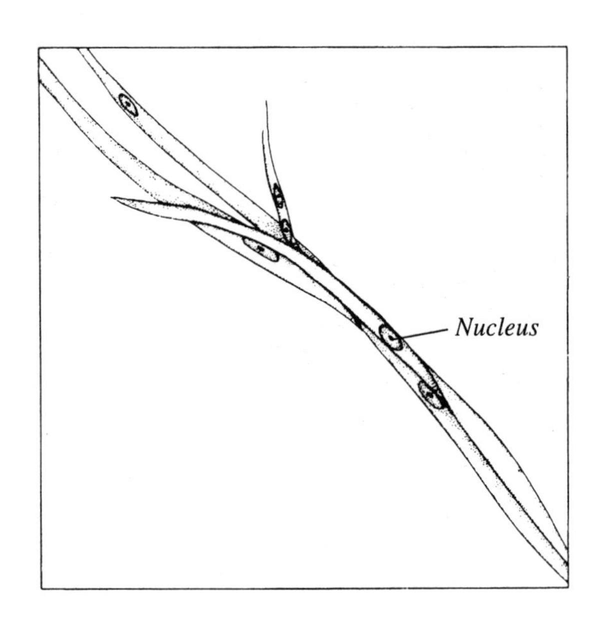

j. *Smooth muscle*

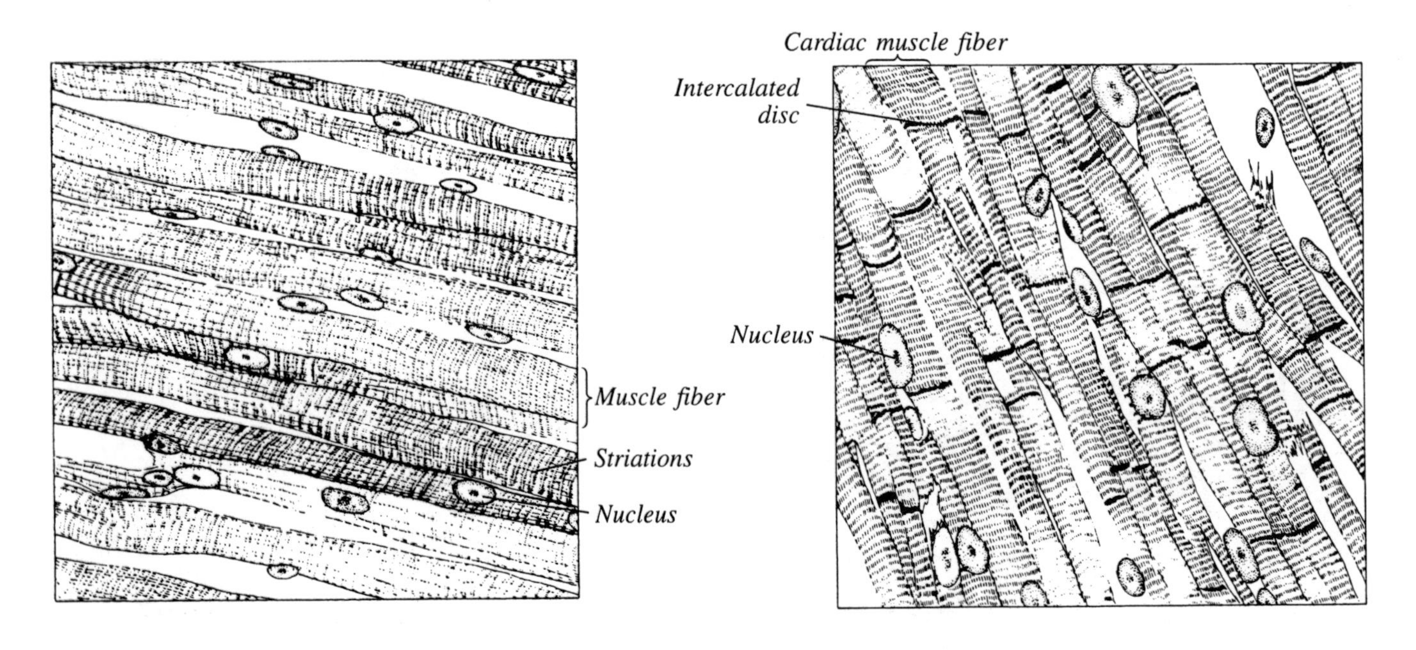

k. *Skeletal muscle*

l. *Cardiac muscle*

NAME ______________________ LAB TIME/DATE ______________

REVIEW SHEET
exercise
6

The Skin (Integumentary System)

Basic Structure of the Skin

1. Complete the following statements by writing the appropriate word or phrase on the correspondingly numbered blank:

The two basic tissues of which the skin is composed are dense connective tissue, which makes up the dermis, and __1__, which forms the epidermis. Most cells of the epidermis are __2__. The protein __3__ makes the dermis tough and leatherlike. The specialized cells that produce the pigments that contribute to skin color are called __4__.

1. *epithelium*
2. *keratinocytes*
3. *collagen*
4. *melanocytes*

2. Four protective functions of the skin are *protection from mechanical damage, chemical damage, thermal damage, and bacterial invasion*

3. Using the key choices, choose all responses that apply to the following descriptions.

Key:

stratum basale	stratum lucidum	reticular layer
stratum corneum	stratum spinosum	epidermis (as a whole)
stratum granulosum	papillary layer	dermis (as a whole)

stratum granulosum 1. layer containing sacs filled with fatty material or keratin subunits

stratum lucidum/stratum corneum 2. dead cells

papillary layer 3. the more superficial dermis layer

epidermis 4. avascular region

dermis 5. major skin area where derivatives (nails and hair) reside

stratum basale 6. epidermal region exhibiting the most mitoses

stratum corneum 7. most superficial epidermal layer

dermis 8. has abundant elastic and collagenic fibers

stratum basale 9. region where melanocytes are most likely to be found

stratum corneum 10. accounts for most of the epidermis

4. Label the skin structures and areas indicated in the accompanying diagram of skin.

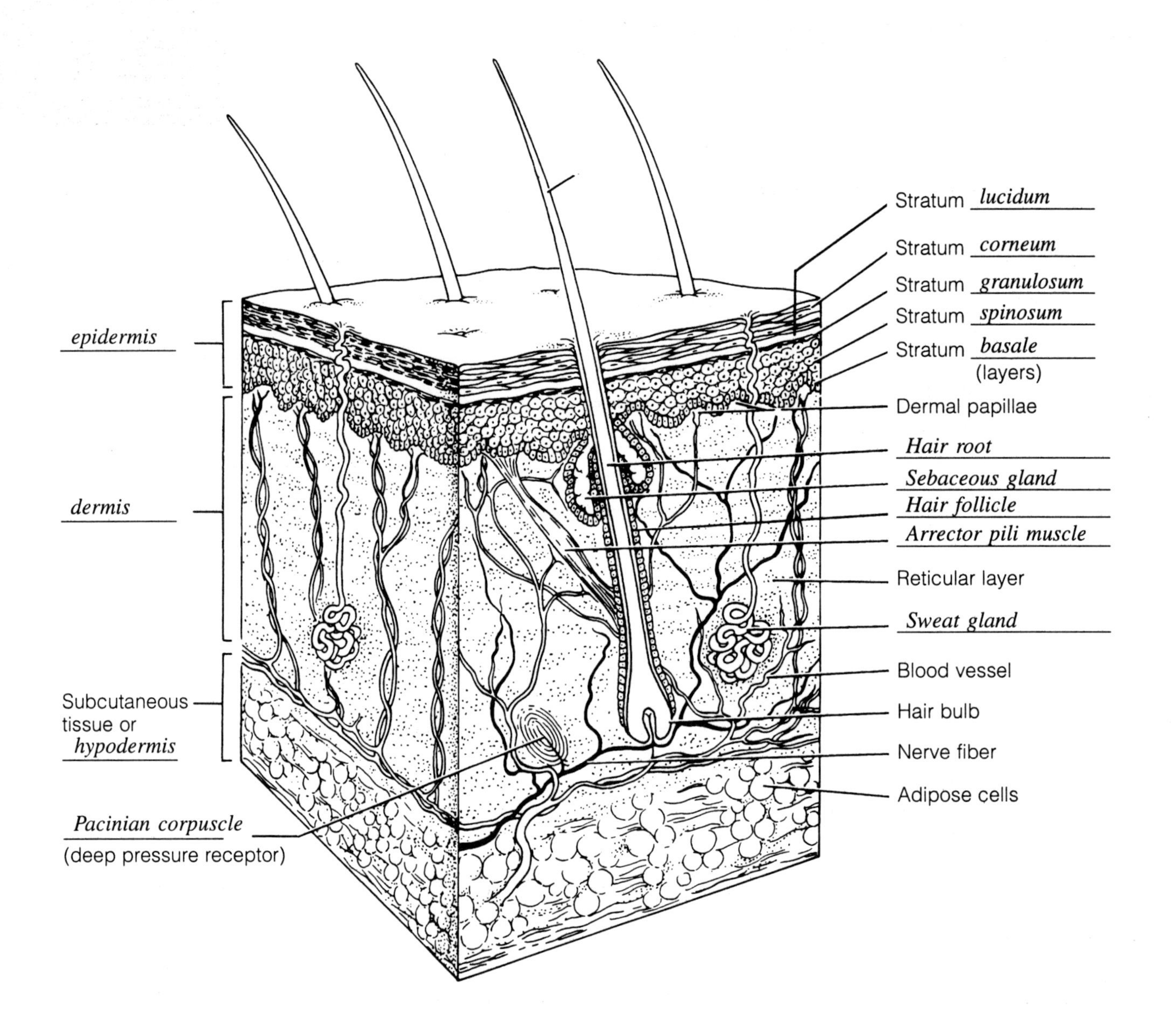

5. What substance is manufactured in the skin (but is not a secretion) to play a role elsewhere in the body?

The skin is the site of vitamin D synthesis for the body.

6. How did the activity "Visualizing Changes in Skin Color Due to Continuous External Pressure" relate to formation of decubitus ulcers? (Use your textbook if necessary.)

Any restriction of the normal blood supply to the skin results in cell death and, if severe or prolonged, will cause decubitus ulcers.

7. Some injections hurt more than others. On the basis of what you have learned about skin structure, can you determine why this is so? *The dermis has a rich nerve supply; some with nerve endings that respond to pain. If these bare nerve endings are stimulated by injection, a pain message will be transmitted to the central nervous system for interpretation.*

8. What was demonstrated by the two-point discrimination test? *The relative density of touch receptors in various body areas (lips, fingertips, etc.)*

9. Two questions regarding general sensation are posed below. Answer each by placing your response in the appropriately numbered blanks to the right.

1–2. Which two body areas tested were most sensitive to touch? 1–2. *lips, fingertips*

3–4. Which two body areas tested were the least sensitive to touch? 3–4. *back of calf, back of neck*

10. Define *adaptation of sensory receptors:* *Decline in receptor sensitivity and stimulation with prolonged unchanging stimuli.*

11. Why is it advantageous to have pain receptors that are sensitive to all vigorous stimuli, whether heat, cold, or pressure?

Because all of these stimuli, if excessive, cause tissue damage.

Pain receptors do not adapt. Why is this important? *Pain is a warning of actual or potential tissue damage.*

12. Imagine yourself without any cutaneous sense organs. Why might this be very dangerous? *Many external stimuli (heat, cold, pressure), which can threaten homeostasis, might go undetected and proper protective measures might not be taken.*

Appendages of the Skin

13. Using the key choices, respond to the following descriptions. (Some choices may be used more than once.)

Key: arrector pili; cutaneous receptors; hair; hair follicle; nail; sebaceous glands; sweat gland—apocrine; sweat gland—eccrine

sebaceous glands 1. Acne is an infection of a(n) ________.

hair follicle 2. Structure that houses a hair.

sweat gland—eccrine 3. More numerous variety of perspiration gland that produces a secretion containing water, salts, and vitamin C; activated by rise in temperature.

hair follicle 4. Sheath formed of both epithelial and connective tissues.

sweat gland—apocrine 5. Type of perspiration-producing gland that produces a secretion containing proteins and fats in addition to water and salts.

sebaceous glands/hair follicle 6. Found everywhere on body except palms of hands and soles of feet.

hair/nail 7. Primarily dead/keratinized cells.

arrector pili 8. Specialized structures that respond to environmental stimuli.

sebaceous glands 9. Its secretion contains cell fragments.

nail 10. "Sports" a lunula and a cuticle.

14. How does the skin help to regulate body temperature? (Describe two different mechanisms.)

1. Capillaries in the papillary layer of the dermis allow heat to radiate to the skin surface to cool off the body and will constrict blood flow to the dermis temporarily when body heat needs to be conserved.

2. Sweat glands secrete perspiration that evaporates and carries large amounts of body heat with it.

15. Several structures or skin regions are lettered in the photomicrograph below. Identify each by matching its letter with the appropriate description that follows.

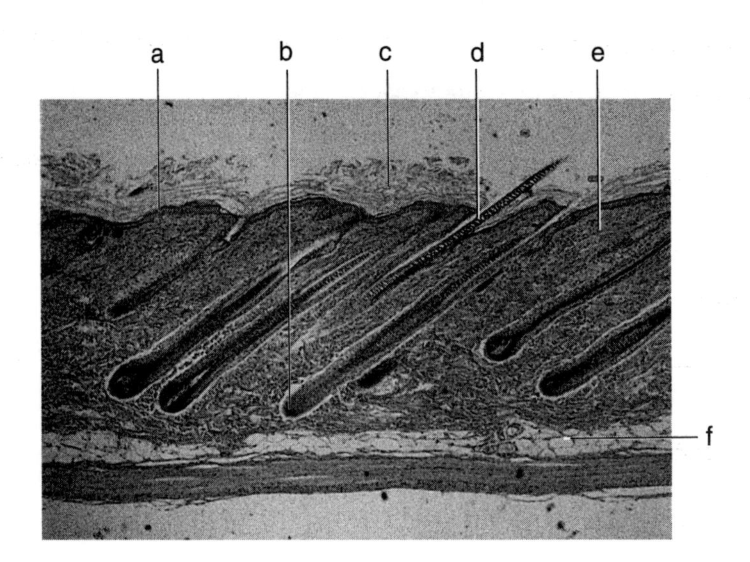

f adipose cells

e dermis

a epidermis

b hair follicle

d hair shaft

c sloughing stratum corneum cells

Plotting the Distribution of Sweat Glands

16. With what substance in the bond paper does the iodine painted on the skin react? *Starch*

17. Which skin area—the forearm or palm of hand—has more sweat glands? *Palm of hand*

Which other body areas would, if tested, prove to have a high density of sweat glands? *Soles of feet, underarms, forehead*

18. What organ system controls the activity of the eccrine sweat glands? *Nervous system*

NAME ______________________ LAB TIME/DATE ______________________

REVIEW SHEET
exercise
7

Overview of the Skeleton

Bone Markings

1. Match the terms in column B with the appropriate description in column A:

	Column A	Column B
spine	1. sharp, slender process	condyle
tubercle	2. small rounded projection	foramen
tuberosity	3. large rounded projection	fossa
head	4. structure supported on neck	head
ramus	5. armlike projection	meatus
condyle	6. rounded, convex projection	ramus
meatus	7. canal-like structure	sinus
foramen	8. opening through a bone	spine
fossa	9. shallow depression	trochanter
sinus	10. air-filled cavity	tubercle
trochanter	11. large, irregularly shaped projection	tuberosity

Classification of Bones

2. The four major anatomical classifications of bones are long, short, flat, and irregular. Which category has the least amount of spongy bone relative to its total volume? *Long bones are mostly compact bone*

3. Classify each of the bones in the chart on the next page into one of the four major categories by checking the appropriate column. Use appropriate references as necessary.

	Long	Short	Flat	Irregular
Humerus	✓			
Phalanx	✓			
Parietal (skull bone)			✓	
Calcaneus (tarsal bone)		✓		
Rib			✓	
Vertebra				✓

Gross Anatomy of the Typical Long Bone

4. Use the terms below to identify the structures marked by leader lines and brackets in the diagrams (some terms are used more than once). After labeling the diagrams, use the listed terms to characterize the statements following the diagrams.

Key:

articular cartilage	epiphyseal line	red marrow cavity
compact bone	epiphysis	trabeculae of spongy bone
diaphysis	medullary cavity	yellow marrow
endosteum	periosteum	

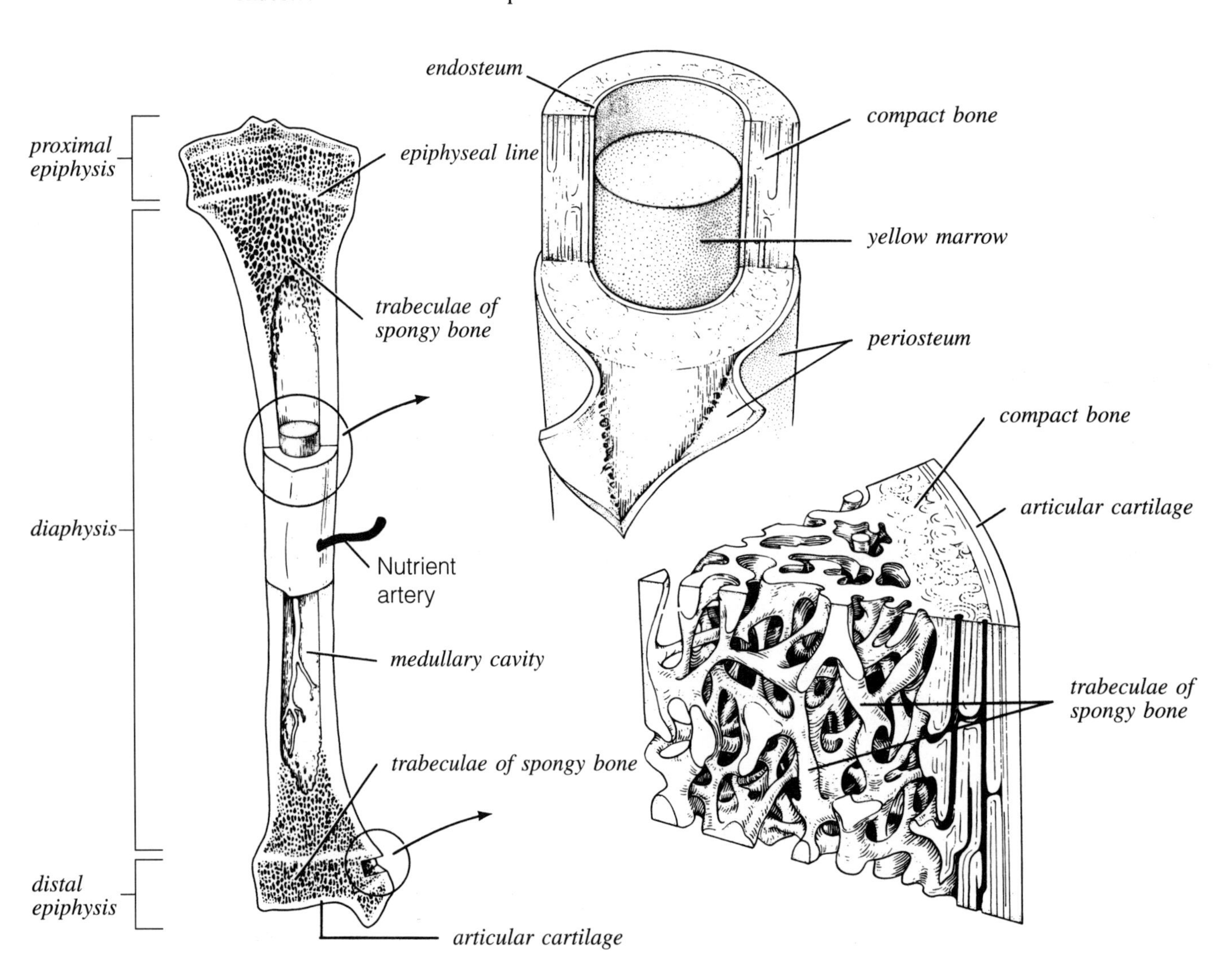

diaphysis 1. made almost entirely of compact bone

red marrow 2. site of blood cell formation

periosteum 3. fibrous membrane that covers the bone

epiphysis 4. scientific term for bone end

yellow marrow 5. contains fat in adult bones

epiphyseal line 6. growth plate remnant

5. What differences between compact and spongy bone can be seen with the naked eye? *Compact bone is dense, smooth and homogenous. Spongy bone is composed of needlelike pieces of bone (trabeculae) and has lots of open space.*

Chemical Composition of Bone

6. What is the function of the organic matrix in bone? *Flexibility and tensile strength*

7. Name the important organic bone components. *Osteocyte and collagen fibers*

8. Calcium salts form the bulk of the inorganic material in bone. What is the function of the calcium salts?

Hardness; compressional strength

9. Which is responsible for bone structure? (circle the appropriate response)

inorganic portion (bone salts) organic portion 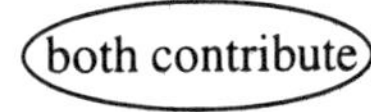 both contribute (circled)

Microscopic Structure of Compact Bone

10. Trace the route taken by nutrients through a bone, starting with the periosteum and ending with an osteocyte in a lacuna.

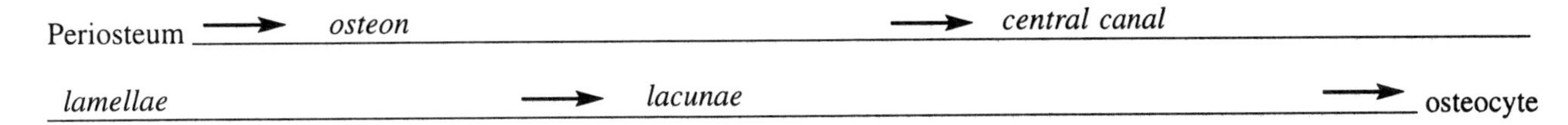

Periosteum → *osteon* → *central canal*

lamellae → *lacunae* → osteocyte

11. On the photomicrograph of bone below (208×), identify all structures listed in the key to the left.

Key: canaliculi
central canal
lamellae (2)
lacuna
bone matrix

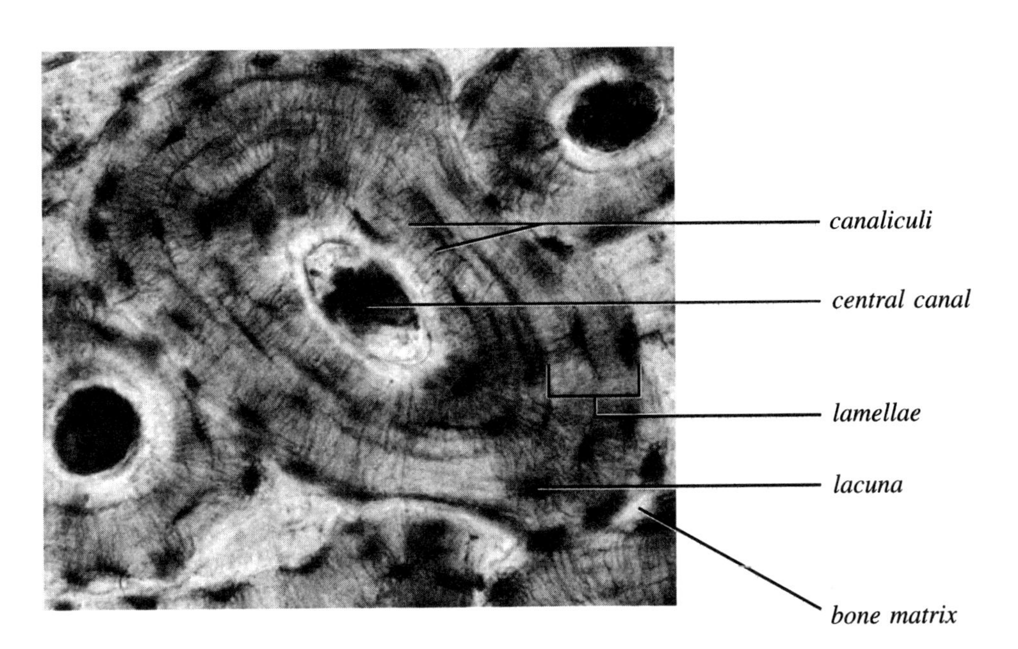

NAME ____________ LAB TIME/DATE ____________

REVIEW SHEET
exercise
8

The Axial Skeleton

The Skull

1. The skull is one of the major components of the axial skeleton. Name the other two.

   vertebrae and *bony thorax*

   What structures do each of these component areas protect? *skull—protects the brain*

   vertebrae—protects the spinal column

   bony thorax—protects thoracic cavity (heart & lungs)

2. Define *suture:* *Interlocking joints; immovable joints that connect bones of skull*

3. With one exception, the skull bones are joined by sutures. Name the exception. *Mandible*

4. What are the four major sutures of the skull, and what bones do they connect?

   Sagittal—two parietal bones

   Coronal—parietals meet frontal bone

   Squamous—temporal meets parietal

   Lambdoid—occipital meets parietal

5. Name the eight bones composing the cranium.

   Frontal *Ethmoid* *Right temporal* *Left temporal*

   Sphenoid *Right parietal* *Left parietal* *Occipital*

6. Give two possible functions of the sinuses.

   Lighten facial bones; act as resonance chambers for speech

7. What is the orbit? *The bony cavity containing the eyeball*

8. Why can the sphenoid bone be called the keystone of the cranial floor? *The sphenoid bone forms a plateau across the width of the skull.*

9. Match the bone names in column B with the descriptions in column A.

	Column A	Column B
frontal	1. bone forming anterior cranium	ethmoid
zygomatic	2. cheekbone	frontal
maxilla	3. upper jaw	hyoid
nasal	4. bony skeleton of nose	lacrimal
palatine	5. posterior roof of mouth	mandible
parietal	6. bone pair united by the sagittal suture	maxilla
temporal	7. site of jugular foramen and carotid canal	nasal
sphenoid	8. contains a "saddle" that houses the pituitary gland	occipital
lacrimal	9. allows tear ducts to pass	palatine
maxilla	10. forms most of hard palate	parietal
ethmoid	11. superior and medial nasal conchae are part of this bone	sphenoid
temporal	12. site of external auditory meatus	temporal
sphenoid	13. has greater and lesser wings	vomer
ethmoid	14. its "holey" plate allows olfactory fibers to pass	zygomatic
maxillary	15. facial bone that contains a sinus	

frontal, *sphenoid*, and *ethmoid* 16. three cranial bones containing paranasal sinuses

occipital	17. its oval-shaped protrusions articulate with the atlas
occipital	18. spinal cord passes through a large opening in this bone
hyoid	19. not really a skull bone
mandible	20. forms the chin
vomer	21. inferior part of nasal septum

mandible, *maxillary* 22. contain alveoli bearing teeth

10. Using choices from column B in question 9 and from the key to the right, identify all bones and bone markings provided with leader lines in the diagram below.

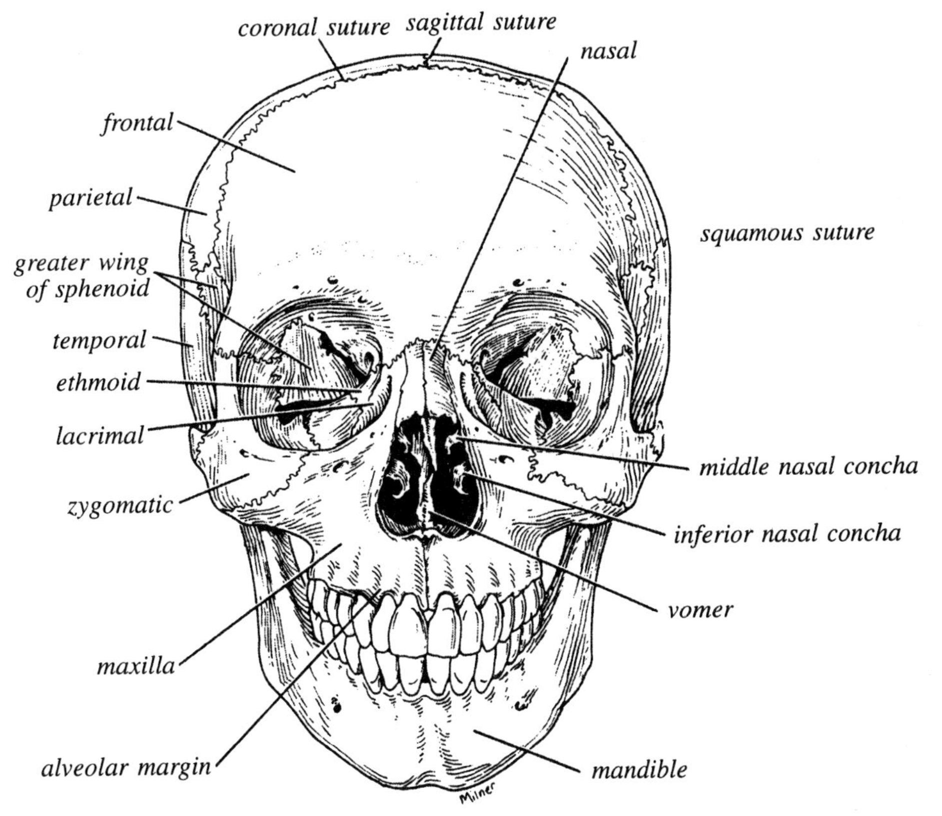

alveolar margin

coronal suture

foramen magnum

greater wing of sphenoid

inferior nasal concha

middle nasal concha of ethmoid

sagittal suture

squamous suture

The Fetal Skull

11. Are the same skull bones seen in the adult also found in the fetal skull? *No, some areas still remain to be converted to bone.*

12. How does the size of the fetal face compare to its cranium? *Face is smaller in proportion to cranium*

How does this compare to the adult skull? *Adult skull is 1/8th total body length whereas the fetal skull is 1/4th total body length.*

13. What are the outward conical projections in some of the fetal cranial bones? *These are growth (ossification) centers.*

14. What is a fontanel? *Fibrous membranes between the bones of a fetal skull*

What is its fate? *Becomes bone by 22 months*

What is the function of the fontanels in the fetal skull? *Allows skull to be compressed during birth and allows for brain growth during late fetal life*

15. Using the terms listed, identify each of the fontanels shown on the fetal skull below.

anterior fontanel | mastoid fontanel | posterior fontanel | sphenoidal fontanel

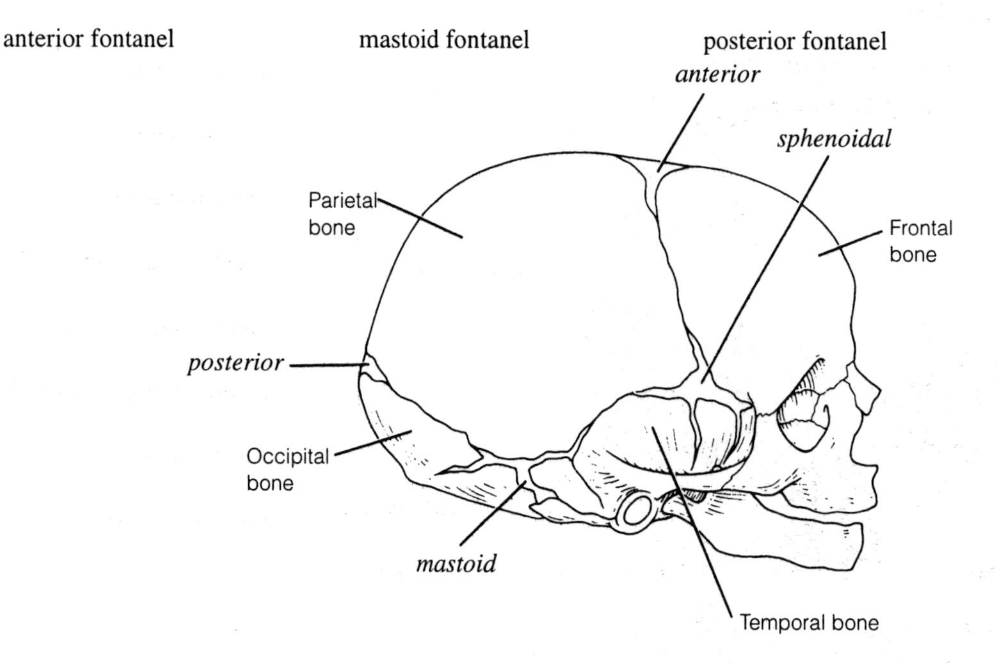

The Vertebral Column

16. Using the key terms, correctly identify the vertebral areas in the diagram.

Key:

body
lamina
pedicle
spinous process
superior articular process
transverse process
vertebral arch
vertebral foramen

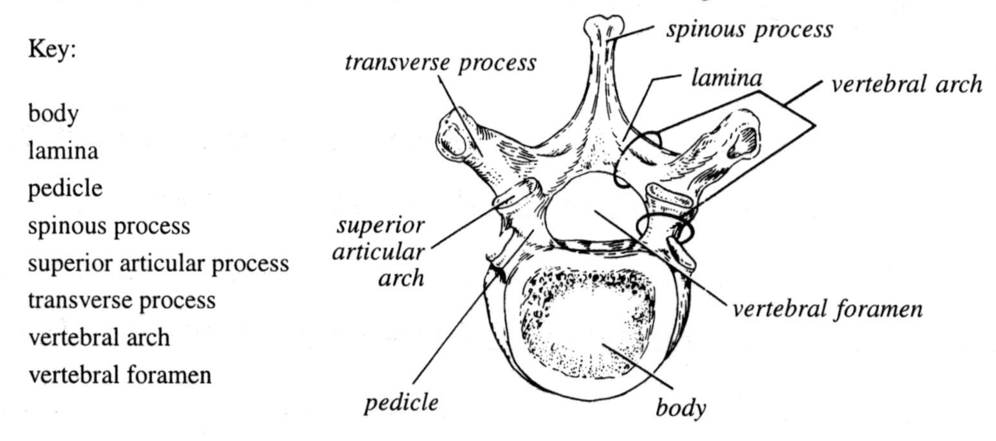

17. The distinguishing characteristics of the vertebrae composing the vertebral column are noted below. Correctly identify each described structure or region by choosing a response from the key.

Key: atlas, axis, cervical vertebra—typical, coccyx, lumbar vertebra, sacrum, thoracic vertebra

cervical vertebra—typical 1. vertebral type with a forked spinous process

atlas 2. pivots on C_2; lacks a body

thoracic vertebra 3. bear facets for articulation with ribs; form part of bony thoracic cage

sacrum 4. forms a joint with the hip bone

lumbar vertebra 5. vertebra with blocklike body and short stout spinous process

coccyx 6. "tail bone"

axis 7. articulates with the occipital condyles

lumbar vertebra 8. five components; unfused

thoracic vertebra 9. twelve components; unfused

sacrum 10. five components; fused

18. Identify as specifically as possible each of the vertebrae types shown in the diagrams below. Also identify and label the following markings on each: transverse processes, spinous process, body, superior articular processes, as well as the areas provided with leaders.

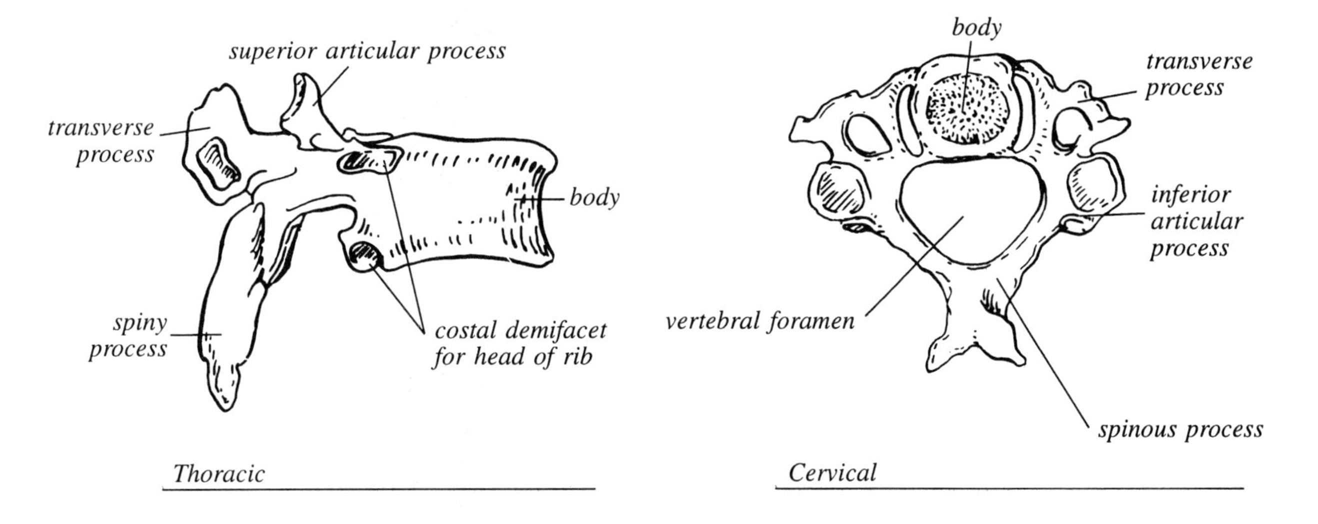

19. What kind of tissue makes up the intervertebral discs? *Fibrocartilage*

20. What is a herniated disc? *A slipped disc; protruding cartilage from vertebra*

What problems might it cause? *Pain and numbness*

21. On this illustration of an articulated vertebral column, identify each structure provided with a leader line by using the key terms.

Key:

atlas

axis

a disc

two thoracic vertebrae

two lumbar vertebrae

sacrum

atlas

axis

thoracic vertebrae

disc

lumbar vertebrae

sacrum

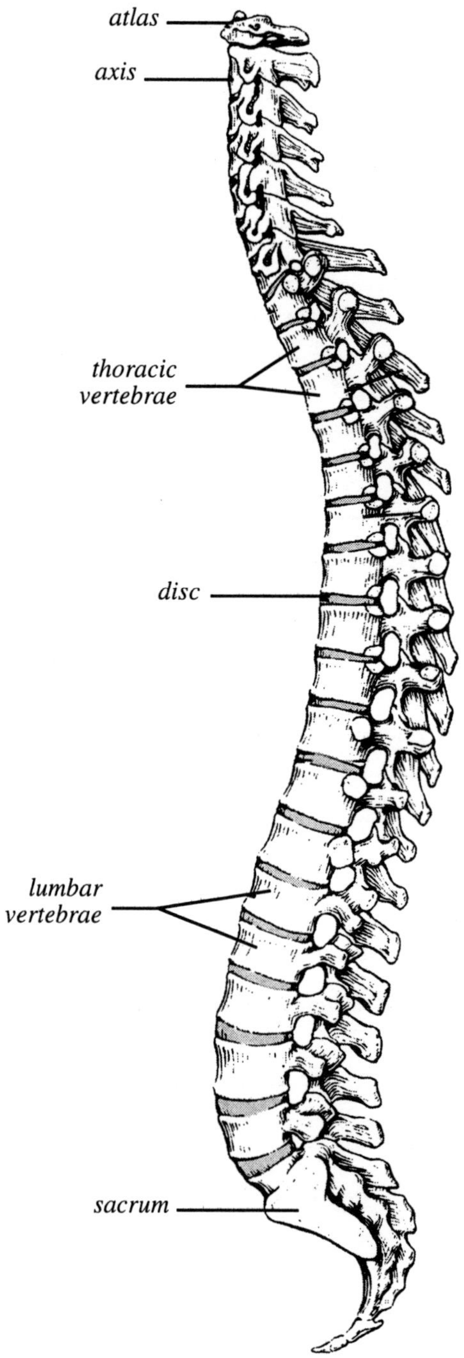

The Bony Thorax

22. The major components of the thorax (excluding the vertebral column) are the *sternum* and the *ribs*.

23. What is the general shape of the thoracic cage? *Cone-shaped*

24. Using the terms at the right, identify the regions and landmarks of the bony thorax.

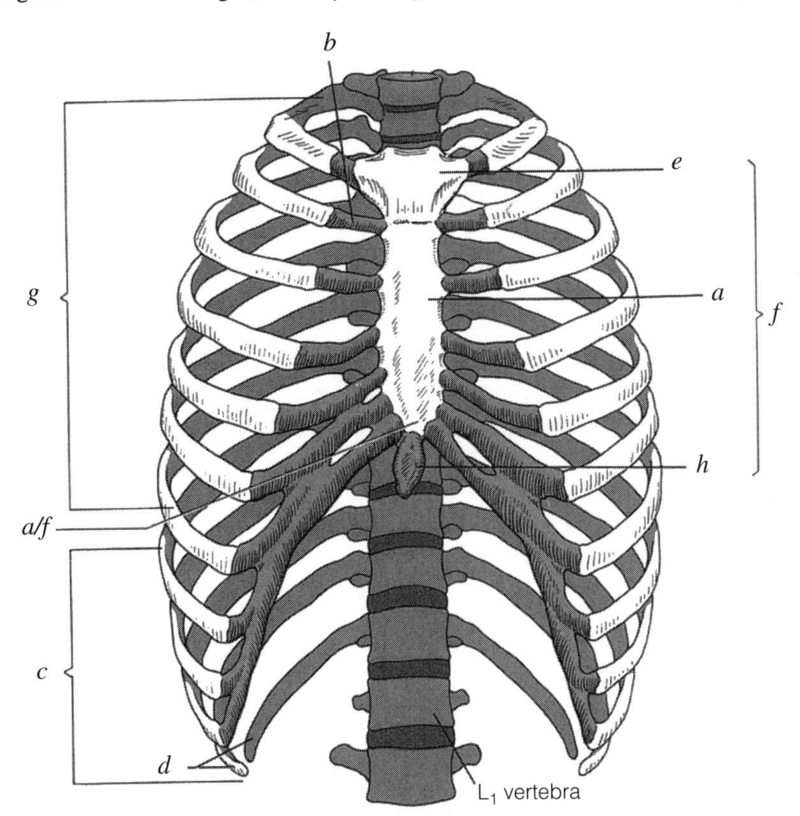

a. body

b. costal cartilage

c. false ribs

d. floating ribs

e. manubrium

f. sternum

g. true ribs

h. xiphoid process

NAME ______________________ LAB TIME/DATE ______________________

REVIEW SHEET
exercise
9

The Appendicular Skeleton

Bones of the Pectoral Girdle and Upper Limb

1. Match the bone names or markings in the key with the leader lines in the figure. The bones are numbered 1–8.

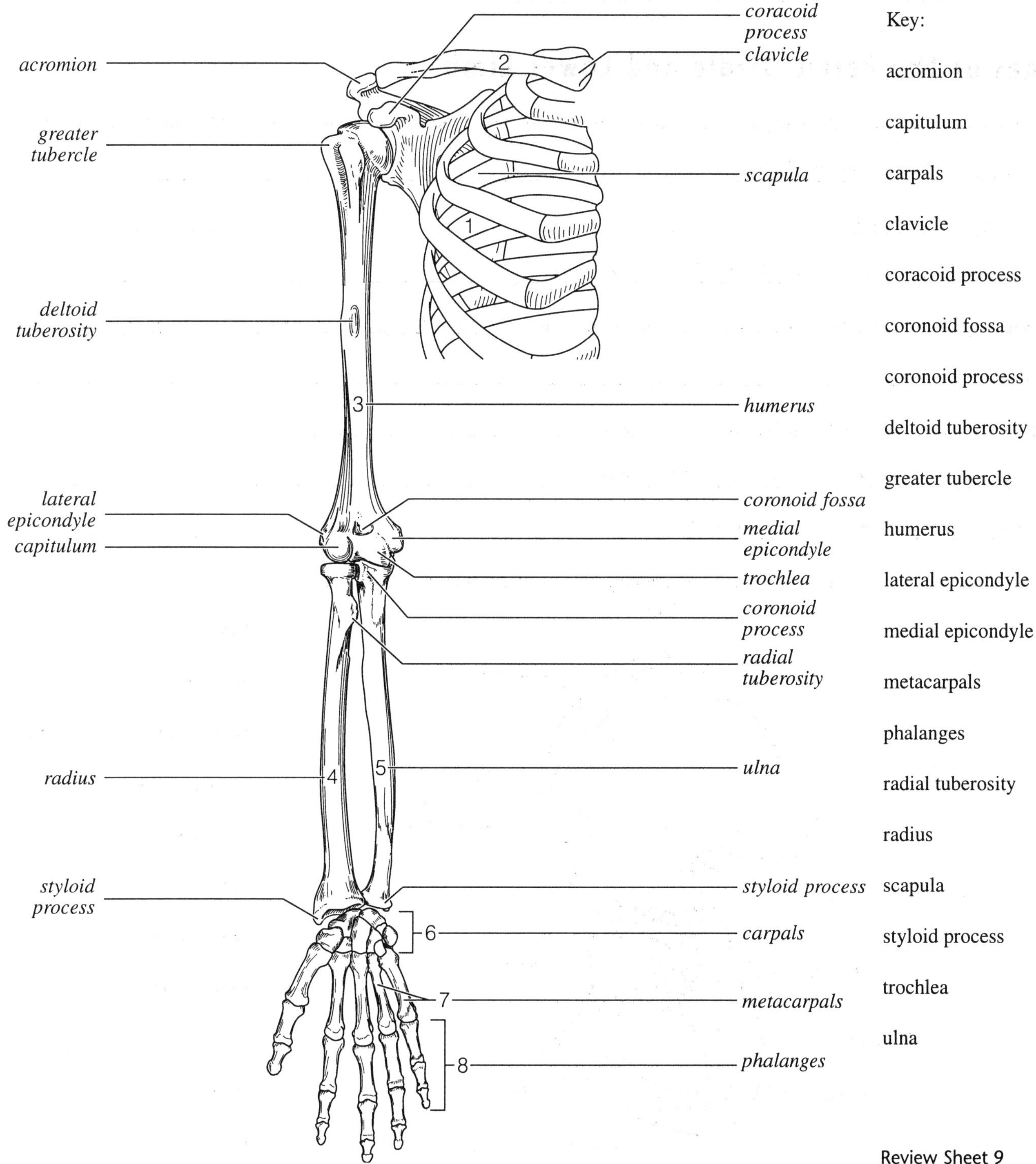

Key:

acromion
capitulum
carpals
clavicle
coracoid process
coronoid fossa
coronoid process
deltoid tuberosity
greater tubercle
humerus
lateral epicondyle
medial epicondyle
metacarpals
phalanges
radial tuberosity
radius
scapula
styloid process
trochlea
ulna

2. Why is the clavicle at risk to fracture when a person falls on his or her shoulder? *It is a slender bone.*

3. Why is there generally no problem in the arm clearing the widest dimension of the thoracic cage?

The clavicle serves as a brace to hold the arm away from the top of the thorax.

4. What is the total number of phalanges in the hand? *14*

5. What is the total number of carpals in the wrist? *8*

Bones of the Pelvic Girdle and Lower Limb

6. Compare the pectoral and pelvic girdles in terms of flexibility (range of motion) allowed, security, and ability to bear weight.

Flexibility: *pectoral—more flexible*

Security: *pelvic—more secure*

Weight-bearing ability: *pelvic—better able to bear weight*

7. What organs are protected, at least in part, by the pelvic girdle? *Reproductive organs, urinary bladder and part of large intestine*

8. Distinguish between the true pelvis and the false pelvis.

false pelvis—superior; supports abdominal viscera

true pelvis—inferior; limits delivery of baby

9. Use terms from the key to identify the bone markings on this illustration of an os coxa.

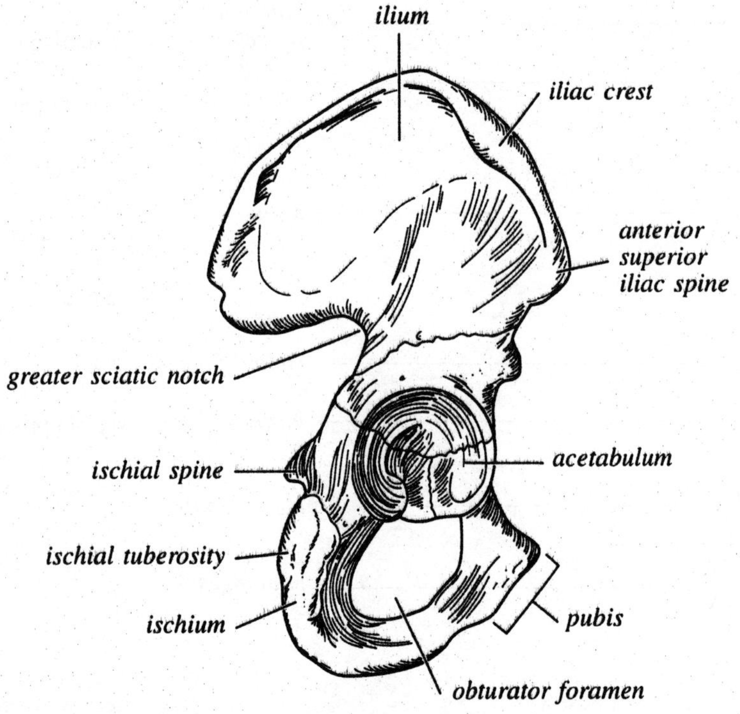

Key:

acetabulum

anterior superior iliac spine

greater sciatic notch

iliac crest

ilium

ischial spine

ischial tuberosity

ischium

obturator foramen

pubis

10. The pelvic bones of a four-legged animal such as the cat or pig are much less massive than those of the human. Make an educated guess as to why this is so.

They do not have to bear as much weight as two-legged animals.

11. A person instinctively curls over the abdominal area in times of danger. Why? *To protect internal organs*

12. What does *fallen arches* mean? *The ligaments and tendons are weakened, allowing bones to "fall."*

13. Match the terms in the key with the appropriate leader lines on the diagram of the femur. Also decide if this bone is a right or left bone.

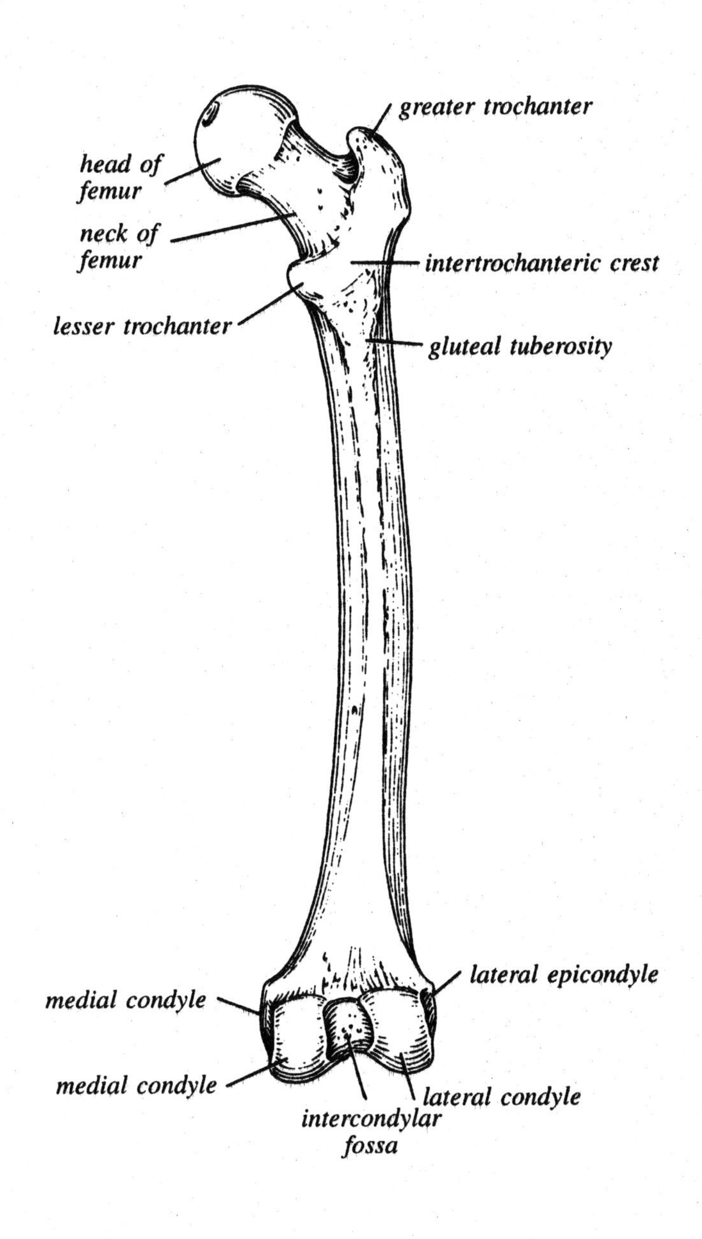

Key:

gluteal tuberosity

greater trochanter

head of femur

intercondylar fossa

intertrochanteric crest

lateral condyle

lateral epicondyle

lesser trochanter

medial condyle

medial epicondyle

neck of femur

The femur shown is the *posterior view of right* member of the two femurs.

14. Match the bone names and markings in the key with the leader lines in the figure. The bones are numbered 1–11.

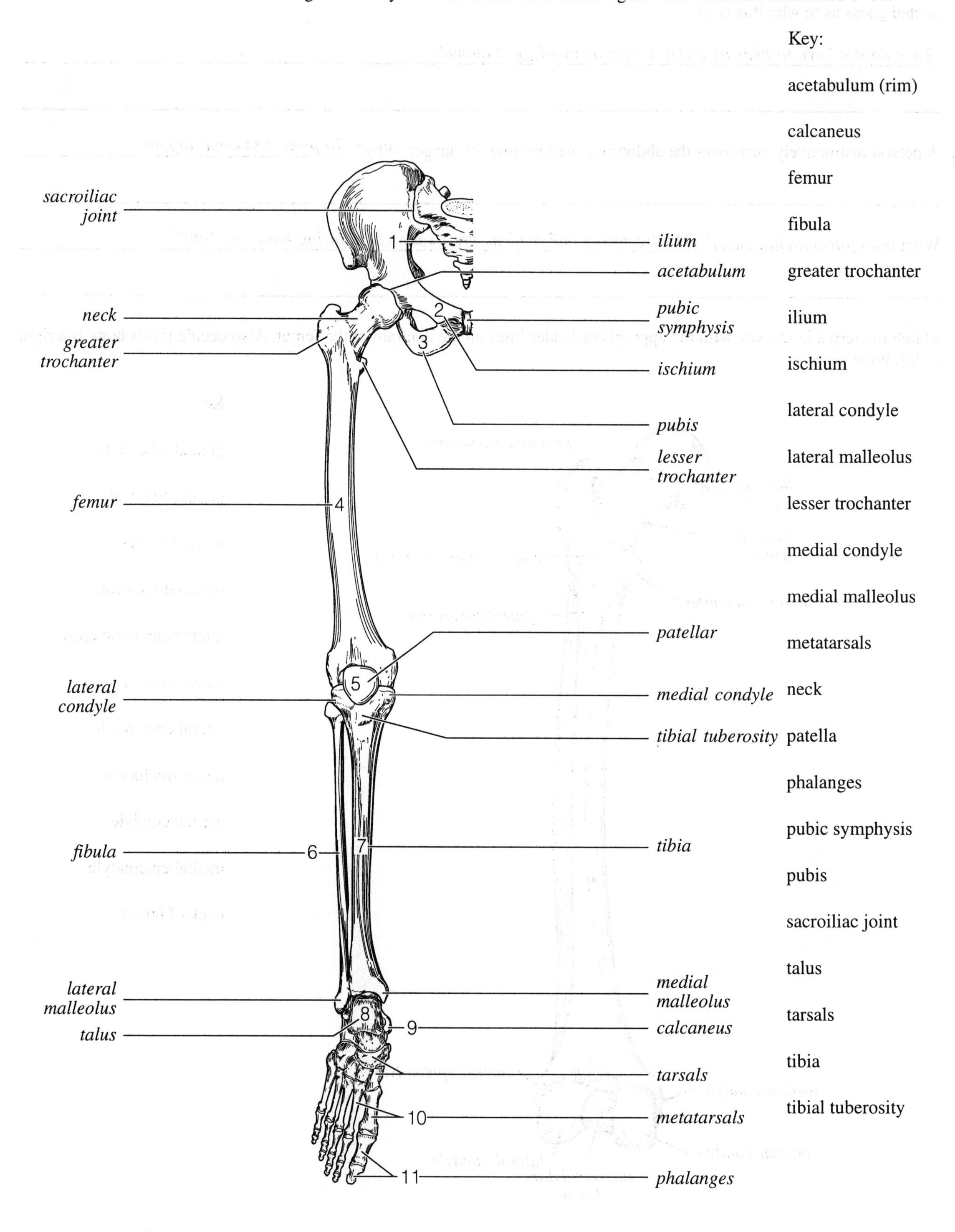

NAME ______________________ LAB TIME/DATE ______________________

REVIEW SHEET

exercise

10

Joints and Body Movements

Types of Joints

1. Use the key terms to identify the joint types described below.

Key: cartilaginous fibrous synovial

cartilaginous 1. typically allows a slight degree of movement

cartilaginous 2. includes joints between the vertebral bodies and the pubic symphysis

fibrous 3. essentially immovable joints

fibrous 4. sutures are the most remembered examples

cartilaginous 5. cartilage connects the bony portions

synovial 6. have a fibrous articular capsule lined with a synovial membrane surrounding a joint cavity

synovial 7. all are freely movable or diarthrotic

fibrous 8. bone regions are united by fibrous connective tissue

synovial 9. include the hip, knee, and elbow joints

2. Match the joint subcategories in column B with their descriptions in column A, and place an asterisk (*) beside all choices that are examples of synovial joints.

	Column A	Column B
suture	1. joint between most skull bones	ball and socket
*pivot**	2. joint between the axis and atlas	condyloid
*ball & socket**	3. hip joint	gliding
*condyloid**	4. joint between forearm bones and wrist	hinge
*hinge**	5. elbow	pivot
*hinge**	6. interphalangeal joints	saddle
*gliding**	7. intercarpal joints	suture
*condyloid**	8. joint between the skull and vertebral column	symphysis
*condyloid**	9. joints between proximal phalanges and metacarpal bones	syndesmosis

3. What characteristics do all joints have in common? *They hold bones together.*

4. Describe the structure and function of the following structures or tissues in relation to a synovial joint and label the structures indicated by leader lines in the diagram.

ligament *reinforce articular capsule*

articular cartilage *covers the ends of the bones*

synovial membrane *lines the articular capsule*

bursa *fluid-filled synovial membrane sacs*

bone
bursa
articular cartilage
synovial membrane
articular capsule
bone
periosteum

5. Which joint, the hip or the knee, is more stable? *hip*

Name two important factors that contribute to the stability of the hip joint.

multiaxial joint and *allows movement in all directions*

Movements Allowed by Synovial Joints

6. Label the *origin* and *insertion* points on the diagram below and complete the following statement:

During muscle contraction, the *insertion* moves toward the *origin*.

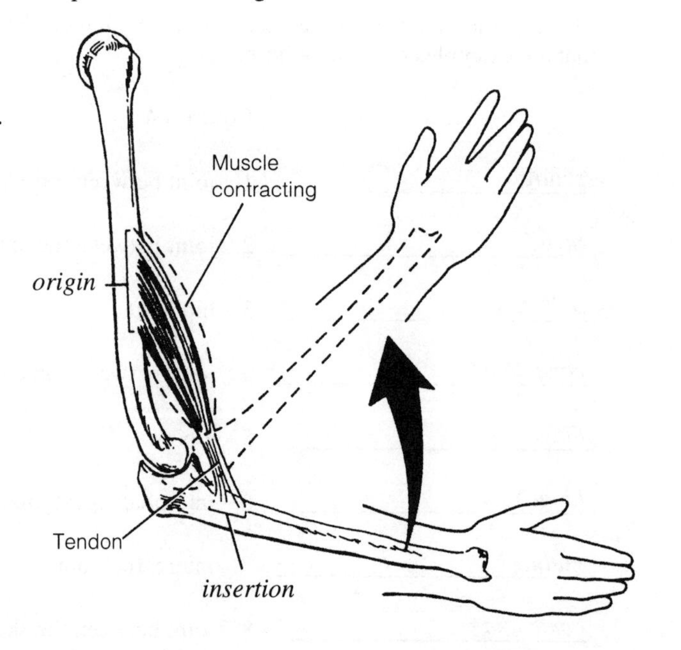

7. Identify the movements demonstrated in the photos by inserting the missing words in the corresponding numbered answered blanks.

1. *flexion*
2. *extension*
3. *hyperextension*
4. *dorsiflexion*
5. *plantar flexion*
6. *rotation*
7. *abduction*
8. *adduction*
9. *circumduction*

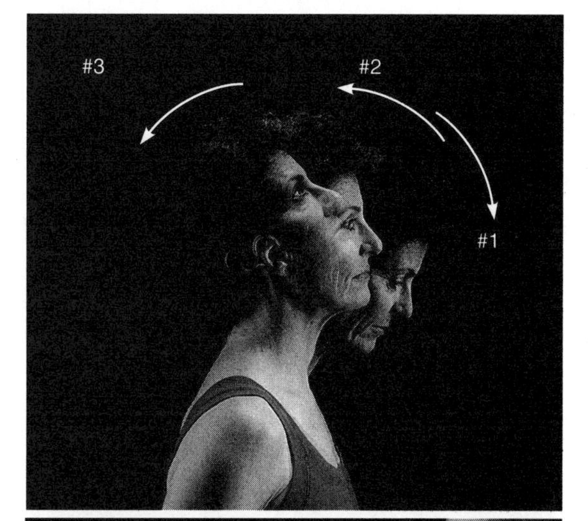

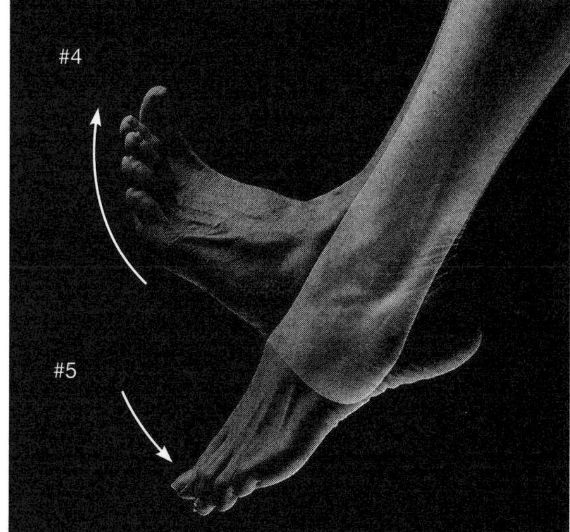

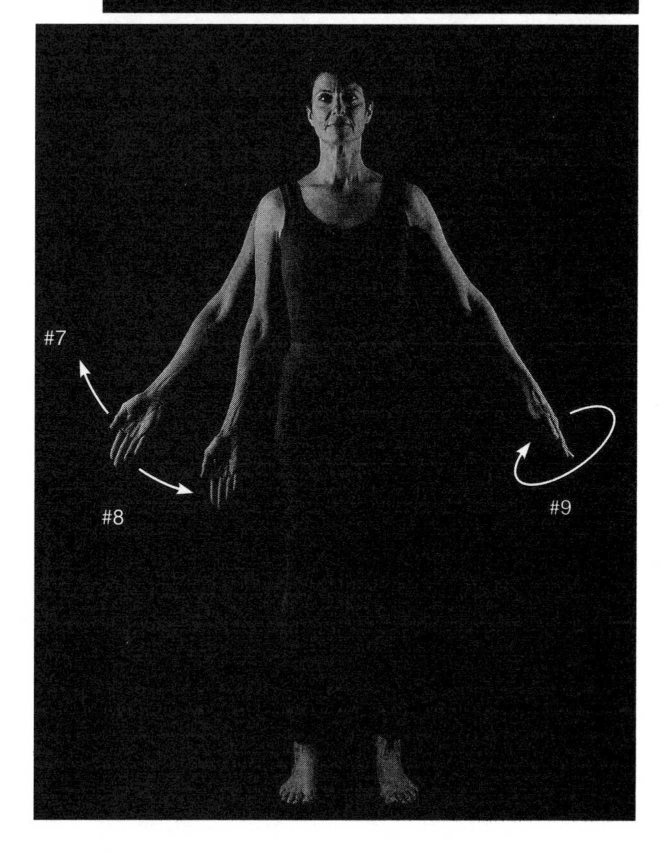

Joint Disorders

8. What structural joint changes are common in older people? *Osteoarthritis; softening, fraying and eventual breakdown of the cartilage leading to bone spurs*

9. Define *dislocation:* *When a bone is forced out of its normal position in the joint cavity*

NAME ______________________ LAB TIME/DATE ______________________

REVIEW SHEET
exercise
11

Microscopic Anatomy and Organization of Skeletal Muscle

Skeletal Muscle Cells and Their Packaging into Muscles

1. From the inside out, name the three types of connective tissue wrappings of a skeletal muscle.

   a. *endomysium* b. *perimysium* c. *epimysium*

   Why are the connective tissue wrappings of skeletal muscle important? (Give at least three reasons.)

   They support and bind muscle fibers, strengthen the muscle as a whole, and provide a route for the entry and exit of nerves and blood vessels that serve the muscle fibers.

2. Why are there more indirect—that is, tendinous—muscle attachments than direct muscle attachments? (Your text may help on this.)

   Tendons provide durability and conserve space. They are tough collagen fibers so they can cross rough, bony projections that would tear delicate muscle tissues. Because of their small size, more tendons can pass over a joint.

3. On the following figure, label endomysium, perimysium, epimysium, and fascicle.

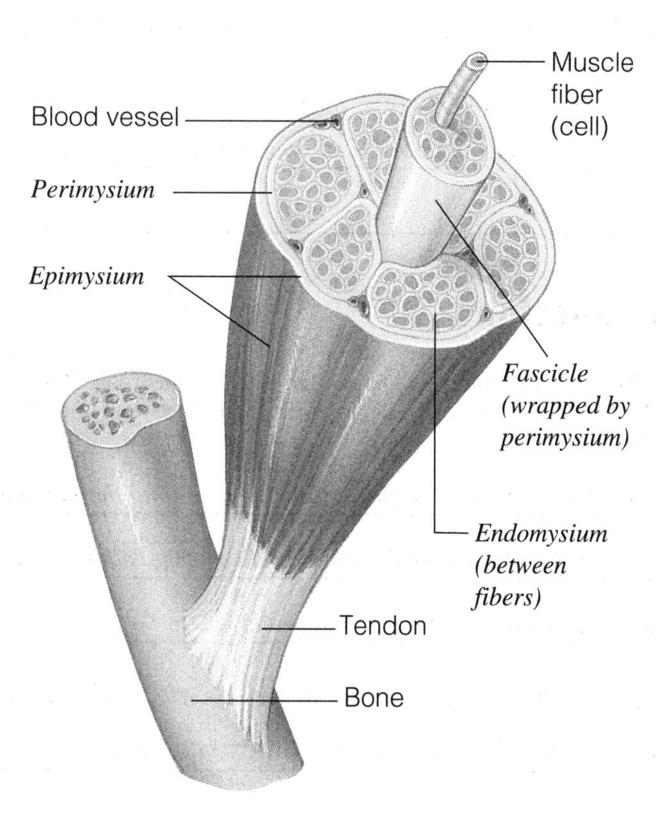

4. The diagram illustrates a small portion of a muscle myofibril in a highly simplified way. Using terms from the key, correctly identify each structure indicated by a leader line or a bracket. Below the diagram make a sketch of how this segment of the myofibril would look if contracted.

Key: actin filament, A band, I band, myosin filament, sarcomere, Z disc

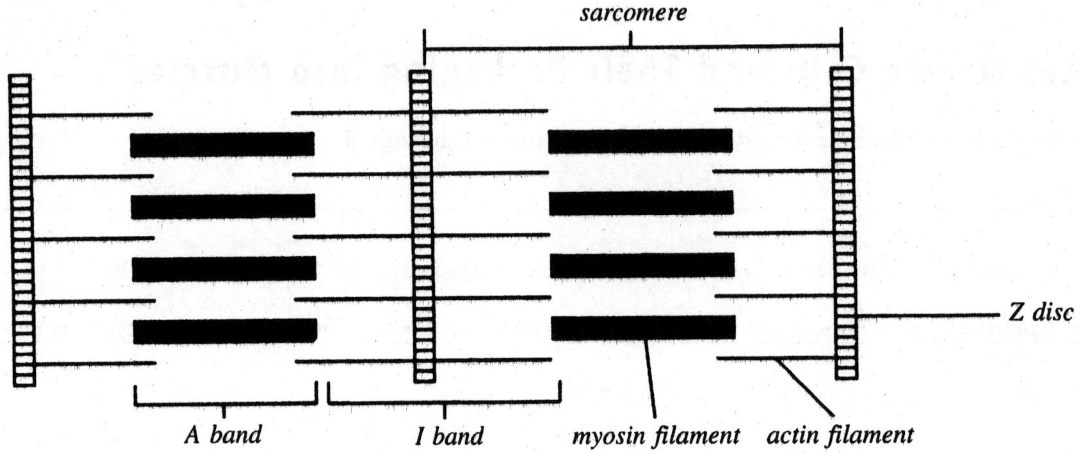

sarcomere

Z disc

A band *I band* *myosin filament* *actin filament*

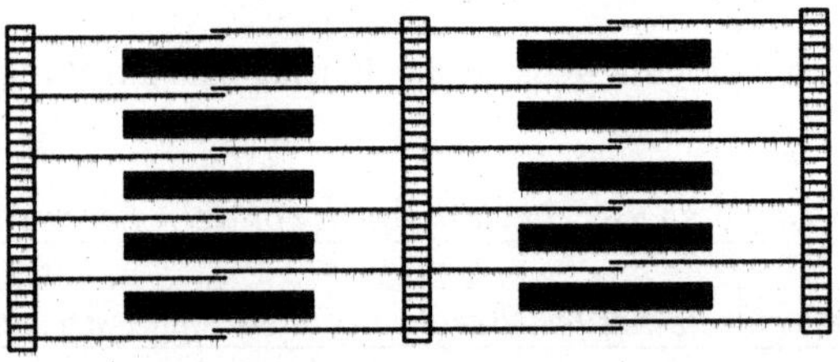

5. Relative to your observations of muscle fiber contraction (pp. 85–86):

a. What percentage of contraction was observed with the solution containing ATP, K^+, and Mg^{2+}? ________%

With *just* ATP? ________% With *just* Mg^{2+} and K^+? ________%

b. *Explain* your observations. *For contraction to occur the proper ions must be present and cellular energy (ATP) must be readily available.*

The Neuromuscular Junction

6. In order for skeletal muscle cells to contract they must be excited by motor neurons. However, the electrical impulse cannot pass directly from a nerve cell to the skeletal muscle cells to excite them. Just what *does pass* from the neuron to the muscle cells, and what effect does it produce?

A neurotransmitter chemical called acetylcholine diffuses from the axon into the synaptic cleft and combines with the receptors on the muscle cells. The permeability of the muscle cells change, allowing more sodium ions to diffuse into the muscle fiber, resulting in the generation of an action potential.

7. Why is it that the electrical impulse cannot pass from neuron to muscle cell? *The neuron and muscle fiber membranes, close as they are, do not actually touch. They are separated by a small fluid-filled gap called the synaptic cleft.*

Classification of Skeletal Muscles

8. Several criteria were given for the naming of muscles. Match the muscle names (column B) to the criteria (column A). Note that more than one muscle may fit the criterion in some cases.

	Column A	Column B
flexor digitorum superficialis	1. action of the muscle	pectoralis major
deltoid	2. shape of the muscle	flexor digitorum superficialis
• biceps brachii • pectoralis major	3. location of the origin and/or insertion of the muscle	biceps brachii
		abdominis transversus
biceps brachii	4. number of origins	erector spinae
• erector spinae • abdominis transversus • pectoralis major • external intercostals	5. location of the muscle relative to a bone or body region	deltoid
• rectus abdominis • abdominis transversus	6. direction in which the muscle fibers run relative to some imaginary line	rectus abdominis
		external intercostals
pectoralis major	7. relative size of the muscle	

9. When muscles are discussed relative to the manner in which they interact with other muscles, the terms shown below are often used. Define each term.

Antagonist: *muscles that oppose or reverse a movement*

Fixator: *specialized synergists that immobilize the origin of a prime mover*

Prime mover: *muscles that are primarily responsible for producing a particular movement*

Synergist: *aid the action of agonists by reducing undesirable/unnecessary movement*

NAME ____________________ LAB TIME/DATE ____________________

REVIEW SHEET
exercise
12

Gross Anatomy of the Muscular System

Muscles of the Head and Neck

1. Using choices from the list at the right, correctly identify the muscles provided with leader lines on the diagram.

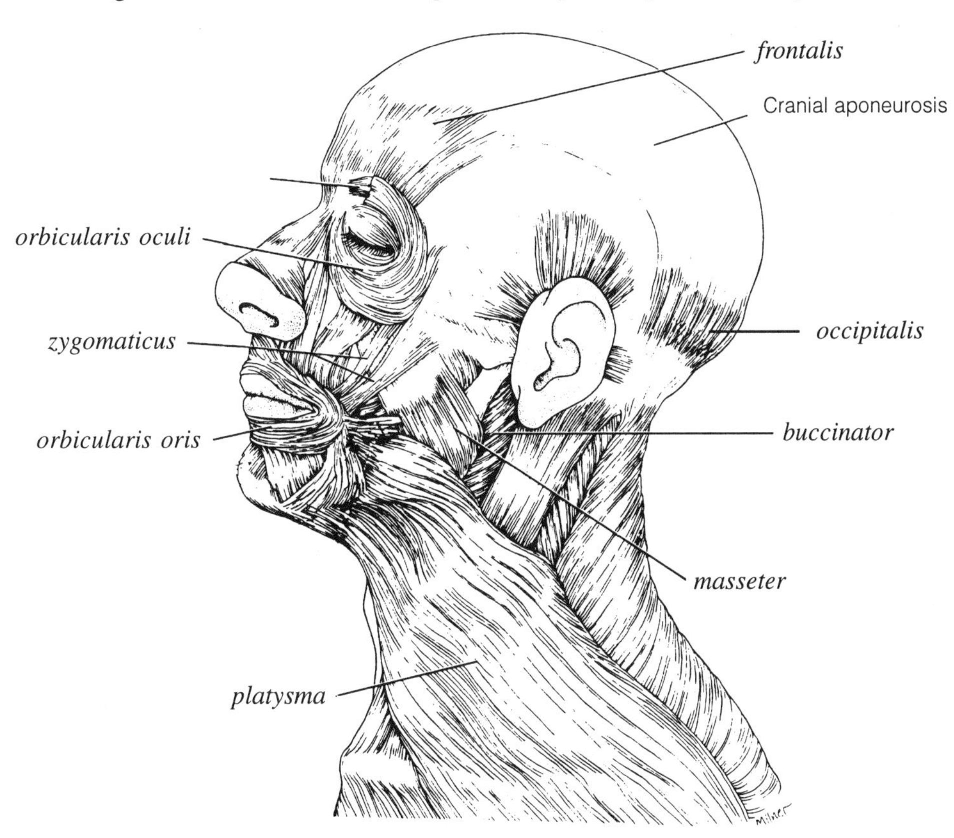

buccinator
frontalis
masseter
platysma
occipitalis
orbicularis oculi
orbicularis oris
zygomaticus

2. Using the terms provided above, identify the muscles described next.

zygomaticus 1. used to grin

buccinator 2. important muscle to a saxophone player

orbicularis oculi 3. used in blinking and squinting

platysma 4. used to pout (pulls the corners of the mouth downward)

frontalis 5. raises your eyebrows for a questioning expression

orbicularis oris 6. your "kisser"

masseter 7. allows you to "bite" that carrot stick

platysma 8. tenses skin of the neck during shaving

Muscles of the Trunk and Upper Limb

3. Using choices from the key, identify the major muscles described next:

rectus abdominis 1. a major spine flexor

latissimus dorsi 2. prime mover for pulling the arm posteriorly

triceps brachii 3. elbow extender

rectus abdominis, *external oblique*, *internal oblique*, *transversus abdominis* 4. help form the abdominal girdle (four pairs of muscles)

extensor carpi ulnaris 5. extends and adducts wrist

deltoid 6. allows you to raise your arm laterally

pectoralis major, *latissimus dorsi* 7. shoulder adductors (two muscles)

biceps brachii 8. flexes elbow; supinates the forearm

external intercostals 9. small muscles between the ribs; elevate the ribs during breathing

erector spinae 10. extends the head

erector spinae 11. extends the spine

extensor carpi radialis 12. extends and abducts the wrist

Key:

biceps brachii
deltoid
erector spinae
extensor carpi radialis
extensor carpi ulnaris
extensor digitorum superficialis
external intercostals
external oblique
flexor carpi radialis
internal oblique
latissimus dorsi
pectoralis major
rectus abdominis
transversus abdominis
trapezius
triceps brachii

Muscles of the Lower Limb

4. Use the key terms to respond to the descriptions below.

Key: adductor group, biceps femoris, gastrocnemius, gluteus maximus, fibularis longus, rectus femoris, semimembranosus, semitendinosus, tibialis anterior, tibialis posterior, vastus muscles

fibularis longus 1. lateral compartment muscle that plantar flexes and everts the ankle

gluteus maximus 2. forms the buttock

gastrocnemius 3. a prime mover of ankle plantar flexion

tibialis anterior 4. a prime mover of ankle dorsiflexion

adductor group 5. allow you to grip a horse's back with your thighs

vastus muscles, *rectus femoris* 6. muscles that insert into the tibial tuberosity (two choices)

rectus femoris 7. muscle that extends knee and flexes thigh

General Review: Muscle Descriptions

5. Identify the muscles described below by completing the statements:.

1. The *deltoid*, *vasti*, and *gluteus maximus and medius*, are commonly used for intramuscular injections (three muscles).
2. The insertion tendon of the *quadriceps* group contains a large sesamoid bone, the patella.
3. The triceps surae insert in common into the *calcaneal* tendon.
4. The bulk of the tissue of a muscle tends to lie *proximal* to the part of the body it causes to move.
5. The extrinsic muscles of the hand originate on the *humerus, radius, and ulna*.
6. Most flexor muscles are located on the *anterior* aspect of the body; most extensors are located *posteriorly*. An exception to this generalization is the extensor-flexor musculature of the *knee*.

General Review: Muscle Recognition

6. Identify the lettered muscles in the diagram of the human anterior superficial musculature by matching the letter with one of the following muscle names:

t 1. orbicularis oris

v 2. pectoralis major

x 3. external oblique

u 4. sternocleidomastoid

g 5. biceps brachii

e 6. deltoid

l 7. vastus lateralis

q 8. frontalis

k 9. rectus femoris

w 10. rectus abdominis

aa 11. sartorius

c 12. platysma

i 13. flexor carpi radialis

r 14. orbicularis oculi

cc 15. gastrocnemius

b 16. masseter

d 17. trapezius

p 18. tibialis anterior

bb 19. adductors

m 20. vastus medialis

z 21. transversus abdominis

n 22. fibularis longus

j 23. iliopsoas

a 24. temporalis

s 25. zygomaticus

f 26. triceps brachii

h 27. brachialis

o 28. extensor digitorum longus

y 29. internal oblique

dd 30. soleus

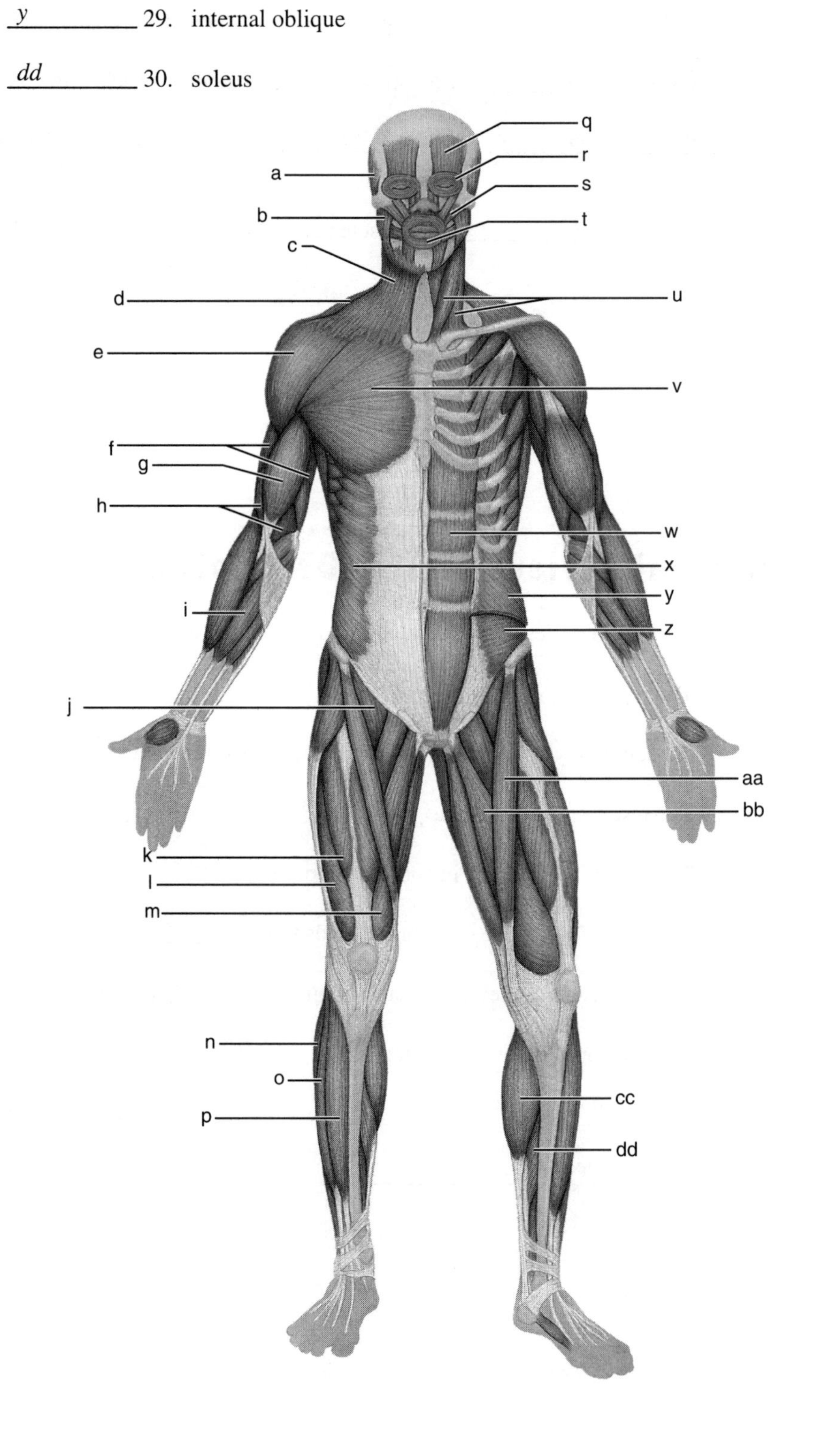

7. Identify each of the lettered muscles in this diagram of the human posterior superficial musculature by matching the letter to one of the following muscle names:

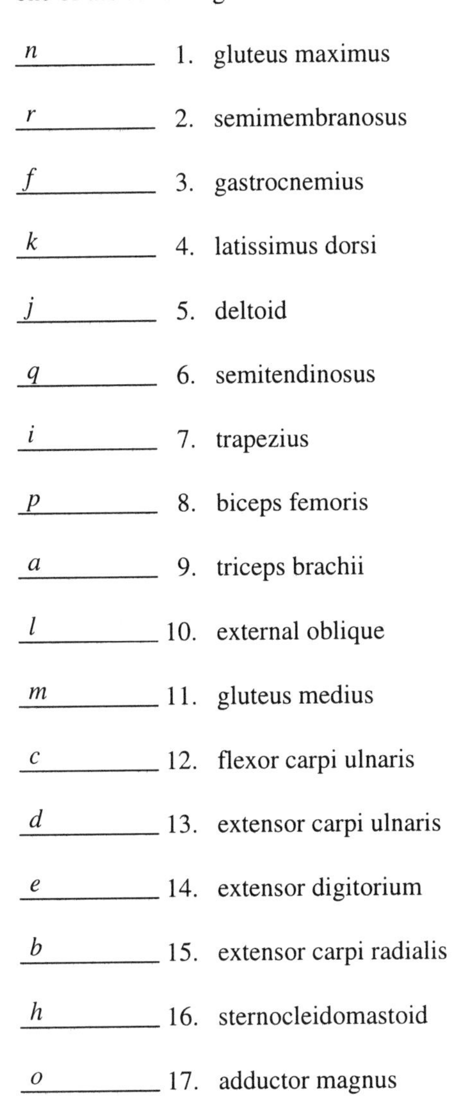

n ________ 1. gluteus maximus

r ________ 2. semimembranosus

f ________ 3. gastrocnemius

k ________ 4. latissimus dorsi

j ________ 5. deltoid

q ________ 6. semitendinosus

i ________ 7. trapezius

p ________ 8. biceps femoris

a ________ 9. triceps brachii

l ________ 10. external oblique

m ________ 11. gluteus medius

c ________ 12. flexor carpi ulnaris

d ________ 13. extensor carpi ulnaris

e ________ 14. extensor digitorium

b ________ 15. extensor carpi radialis

h ________ 16. sternocleidomastoid

o ________ 17. adductor magnus

g ________ 18. soleus

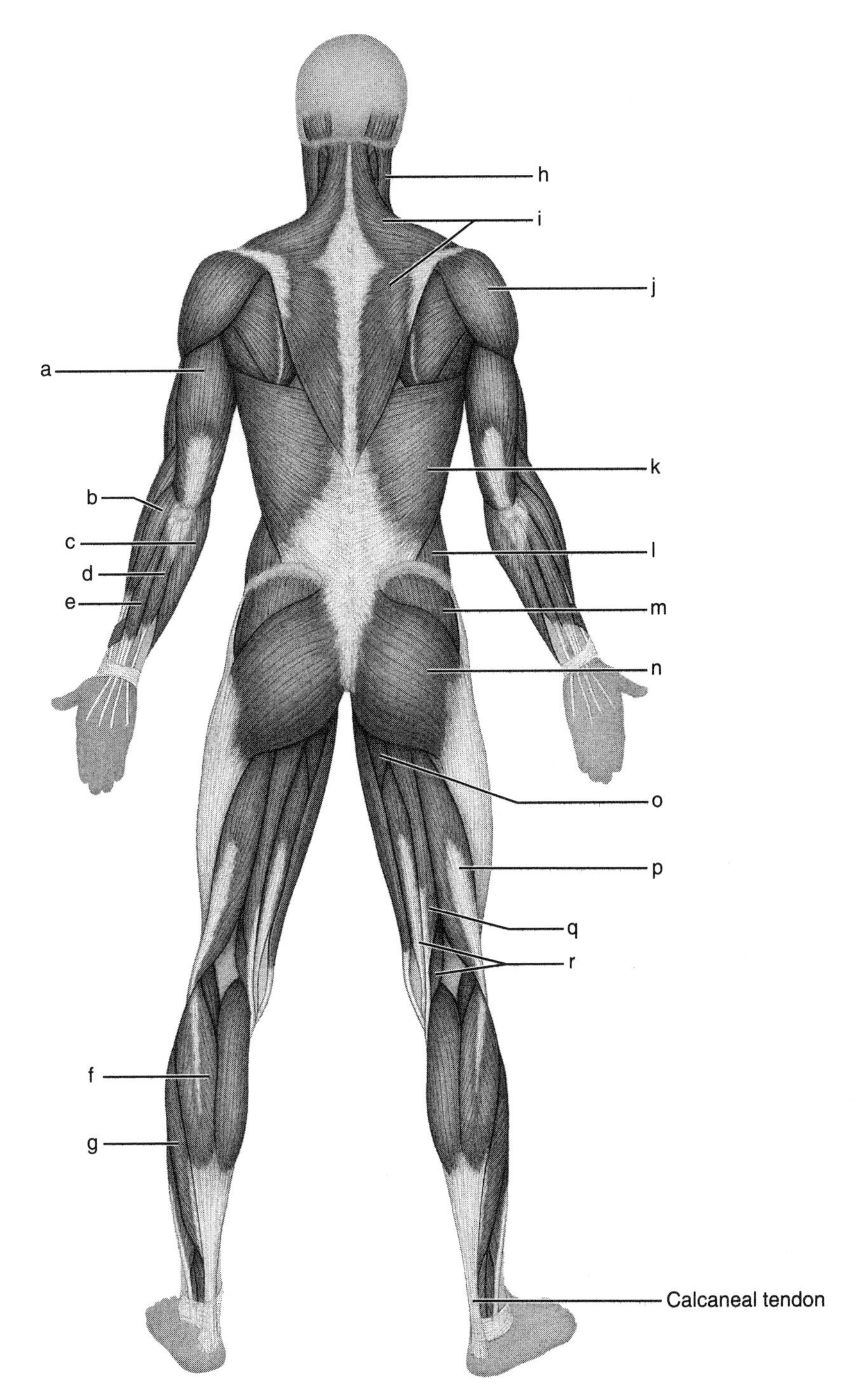

NAME ______________________ LAB TIME/DATE ______________________

REVIEW SHEET
exercise 13

Neuron Anatomy and Physiology

1. The cellular unit of the nervous system is the neuron. What is the major function of this cell type?

The major function of the neuron is to transmit messages (nerve impulses) from one part of the body to another.

2. The supporting cells, or neuroglia, have numerous functions. Name three.

The supporting cells act as phagocytes, protect and myelinate, and act as a selective barrier between the capillary blood supply and the neurons.

3. Match each statement with a response chosen from the key.

Key:			
	afferent neuron	ganglion	peripheral nervous system
	association neuron	neurotransmitters	synapse
	central nervous system	nerve	tract
	efferent neuron	nuclei	

central nervous system 1. the brain and spinal cord collectively

synapse 2. junction or point of close contact between neurons

ganglion 3. a bundle of nerve processes outside the central nervous system

association neuron 4. neuron connecting sensory and motor neurons

tract 5. spinal and cranial nerves and ganglia

nuclei 6. collections of nerve cell bodies inside the CNS

efferent neuron 7. neuron that conducts impulses away from the CNS to muscles and glands

afferent neuron 8. neuron that conducts impulses toward the CNS from the body periphery

neurotransmitters 9. chemicals released by axonal terminals

Neuron Anatomy

4. Draw a "typical" neuron in the space below. Include and label the following structures on your diagram: cell body, nucleus, dendrites, axon, myelin sheath, and nodes of Ranvier.

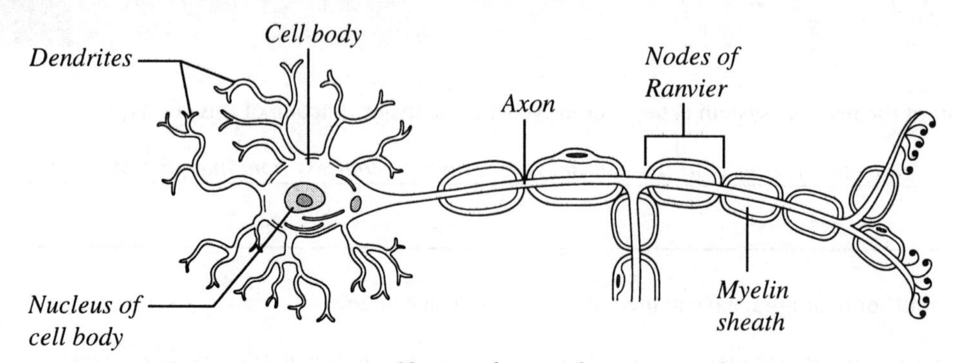

5. How is one-way conduction at synapses ensured? *Neurons have only one axon that carries impulses away from the nerve cell body toward the synapse.*

6. What anatomical characteristic determines whether a particular neuron is classified as unipolar, bipolar, or multipolar?

The number of processes attached to the cell body determines the structural class of a neuron.

Make a simple line drawing of each type here.

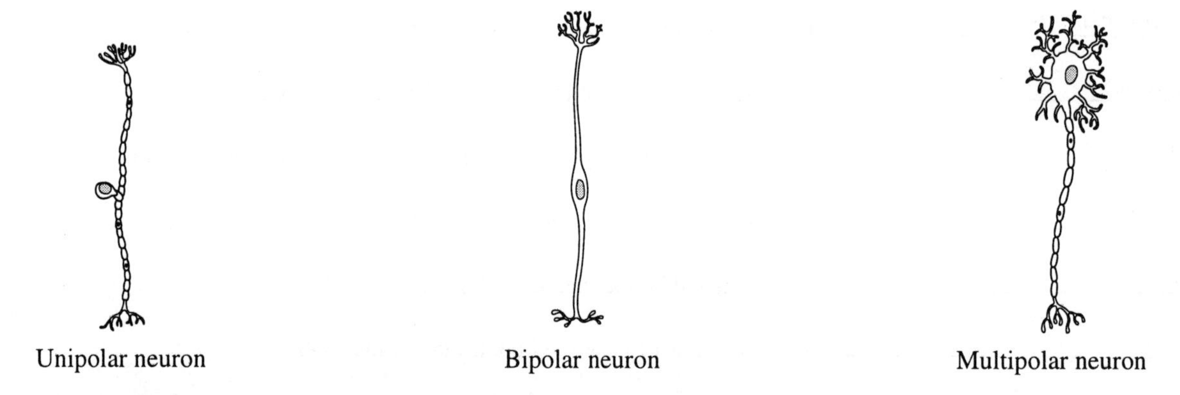

7. Describe how the Schwann cells form the myelin sheath and the neurilemma encasing the nerve processes. (You may want to diagram the process.)

Axons in the peripheral nervous system are myelinated by special supporting cells called Schwann cells, which wrap themselves tightly around the axon in jelly-roll fashion so that when the process is completed, a tight core of plasma membrane material called the myelin sheath encompasses the axon. The Schwann cell nucleus and the bulk of its cytoplasm end up just beneath the outermost portion of its plasma membrane. This part of the Schwann cell which is external to the myelin sheath, is referred to as the neurilemma.

8. Correctly identify the sensory (afferent) neuron, association neuron (interneuron), and motor (efferent) neuron in the figure below.

Which of these neuron types is/are unipolar? *Sensory neuron*

Which is/are most likely multipolar? *Interneuron and motor neuron*

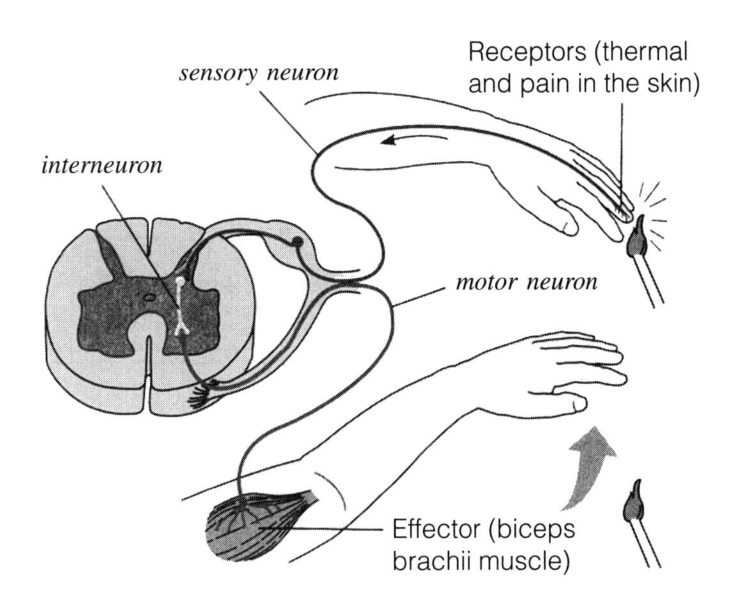

The Nerve Impulse

9. Match each of the terms in column B to the appropriate definition in column A.

	Column A	Column B
depolarization	1. reversal of the resting potential owing to an influx of sodium ions	action potential
repolarization	2. period during which potassium ions are diffusing out of the neuron	depolarization
action potential	3. transmission of the depolarization wave along the neuronal membrane	repolarization
sodium-potassium pump	4. mechanism that restores the resting membrane voltage and intracellular ionic concentrations	sodium-potassium pump

10. Would a substance that decreases membrane permeability to sodium increase *or* decrease the probability of generating a nerve impulse? *It would decrease the probability.*

11. Why don't the terms *depolarization* and *action potential* mean the same thing? (*Hint:* under what conditions will a local depolarization *not* lead to the action potential? *If the stimulus is of less than threshold intensity, depolarization is limited to a small area of the membrane, and no action potential is generated.*

Structure of a Nerve

12. What is a nerve? *A bundle of neuron fibers or processes that extends to and/or from the CNS and visceral organs or structures of the body periphery such as skeletal muscles, glands, and skin.*

13. State the location of each of the following connective tissue coverings:

endoneurium *Surrounds each nerve fiber*

perineurium *Surrounds a group of nerve fibers*

epineurium *Surrounds the bundles of fibers called fascicles*

14. What is the value of the connective tissue wrappings found in a nerve? *The connective tissue wrappings help insulate the nerve.*

15. Define *mixed nerve:* *Nerves carrying both sensory (afferent) and motor (efferent) fibers*

NAME ______________________ LAB TIME/DATE ______________________

REVIEW SHEET

exercise

14

Gross Anatomy of the Brain and Cranial Nerves

The Human Brain

1. In which of the cerebral lobes (frontal, parietal, occipital, or temporal) would the following functional areas be found?

auditory area *Temporal*

olfactory area *Temporal*

primary motor area *Frontal*

visual area *Occipital*

somatic sensory area *Parietal*

Broca's area *Frontal*

2. Match the letters on the diagram of the human brain (right lateral view) to the appropriate terms listed at the left:

h 1. frontal lobe

b 2. parietal lobe

j 3. temporal lobe

f 4. precentral gyrus

c 5. parieto-occipital sulcus

a 6. postcentral gyrus

i 7. lateral sulcus

g 8. central sulcus

e 9. cerebellum

l 10. medulla

d 11. occipital lobe

k 12. pons

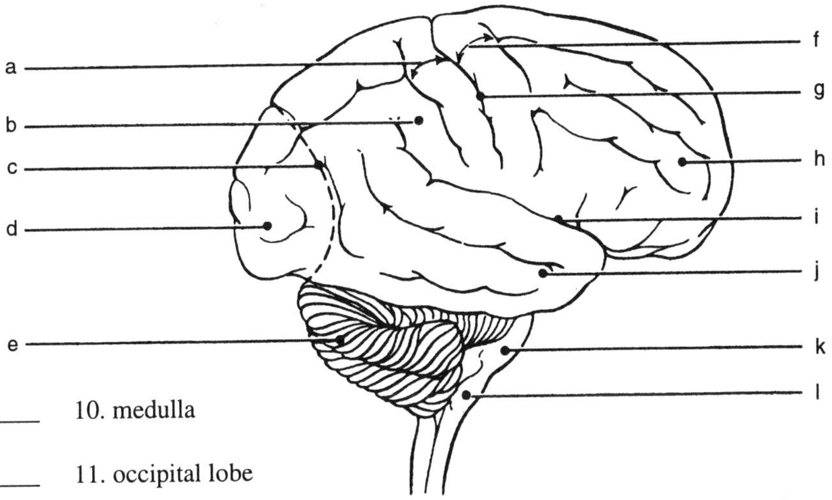

a
b
c
d
e
f
g
h
i
j
k
l

3. Which of the following structures are not part of the brain stem? (Circle the appropriate response or responses.)

(cerebral hemispheres) pons midbrain (cerebellum) medulla

4. Complete the following statements by writing the proper word or phrase in the corresponding blank at the right.

A(n) __1__ is an elevated ridge of cerebral tissue. Inward folds of cerebral tissue are called __2__ or __3__. Gray matter is composed of __4__. White matter is composed of __5__. A fiber tract that provides for communication between different parts of the CNS is called a(n) __6__, whereas one that carries impulses between the periphery and CNS areas is called a(n) __7__. Nuclei deep within the cerebral hemisphere white matter are collectively called the __8__.

1. *gyri*
2. *fissures*
3. *sulci*
4. *cell bodies of neurons*
5. *fiber tracts*
6. *process*
7. *nerves*
8. *basal nuclei (ganglia)*

5. Identify the structures on the following sagittal view of the human brain by matching the lettered areas to the proper terms at the left:

p	1. cerebellum
m	2. cerebral aqueduct
a	3. cerebral hemisphere
l	4. cerebral peduncle
h	5. choroid plexus
k	6. corpora quadrigemina
b	7. corpus callosum
n	8. fourth ventricle
d	9. hypothalamus
f	10. mammillary bodies
c	11. intermediate mass
q	12. medulla oblongata
e	13. optic chiasma
j	14. pineal body
g	15. pituitary gland
o	16. pons
i	17. thalamus

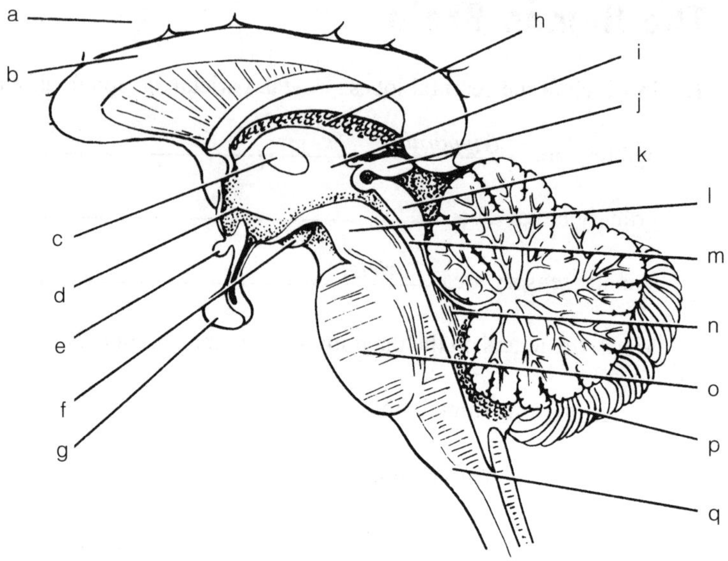

6. Using the anatomical terms from item 4, match the appropriate structures with the following descriptions:

medulla oblongata	1. most important autonomic center of brain
corpora quadrigemina	2. located in the midbrain; contains reflex centers for vision and hearing
cerebellum	3. coordinates complex muscular movements
medulla oblongata	4. contains autonomic centers regulating heart rate, respiration, and other visceral activities
corpus callosum	5. large fiber tract connecting the cerebral hemispheres
pituitary gland; pineal body	6. part of the endocrine system
cerebral aqueduct	7. canal that connects the third and fourth ventricles
thalamus	8. the intermediate mass is part of it

7. Explain why trauma to the base of the brain is often much more dangerous than trauma to the frontal lobes. (*Hint:* Think about the relative function of the cerebral hemispheres and the brain stem structures. Which contain centers more vital to life?)

Base of brain houses brain stem, which houses most of the vital autonomic centers controls → heart rate,

respiration, blood pressure

Meninges of the Brain

8. Identify the meningeal (or associated) structures described below:

dura mater 1. outermost layer; tough fibrous connective tissue

pia mater 2. innermost vascular layer covering the brain; follows every convolution

arachnoid villi 3. drains cerebrospinal fluid into the venous blood in the dural sinuses

choroid plexus 4. structure that forms the cerebrospinal fluid

arachnoid mater 5. middle layer; delicate with cottony fibers

falx cerebri 6. a dural fold that attaches the cerebrum to the crista galli of the skull

Cerebrospinal Fluid

9. Fill in the following flowchart to indicate the path of cerebrospinal fluid from its formation site (assume that this is one of the lateral ventricles) to where it is reabsorbed into the venous blood:

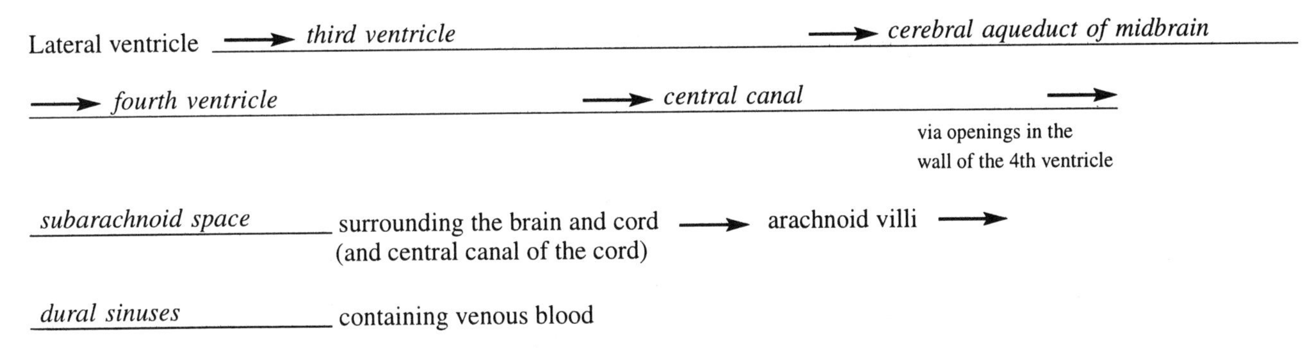

10. Label correctly the structures involved with circulation of cerebrospinal fluid on the accompanying diagram. (These structures are identified by leader lines.)

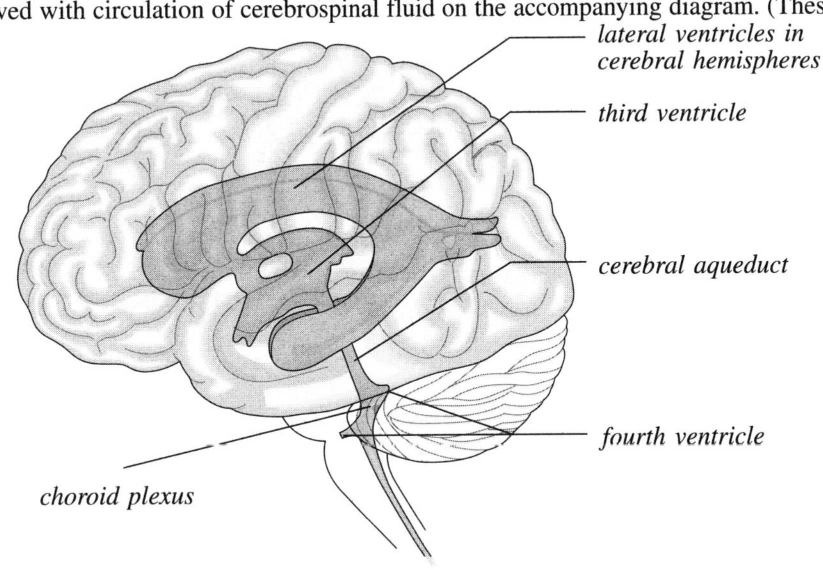

Cranial Nerves

11. Using the following terms, correctly identify all structures indicated by leader lines on the diagram.

abducens nerve (VI)	longitudinal fissure	pituitary gland
accessory nerve (XI)	mammillary body	pons
cerebellum	medulla oblongata	spinal cord
cerebral peduncle	oculomotor nerve (III)	temporal lobe of cerebral hemisphere
facial nerve (VII)	olfactory bulb	trigeminal nerve (V)
frontal lobe of cerebral hemisphere	optic chiasma	trochlear nerve (IV)
glossopharyngeal nerve (IX)	optic nerve (II)	vagus nerve (X)
hypoglossal nerve (XII)	optic tract	vestibulocochlear nerve (VIII)

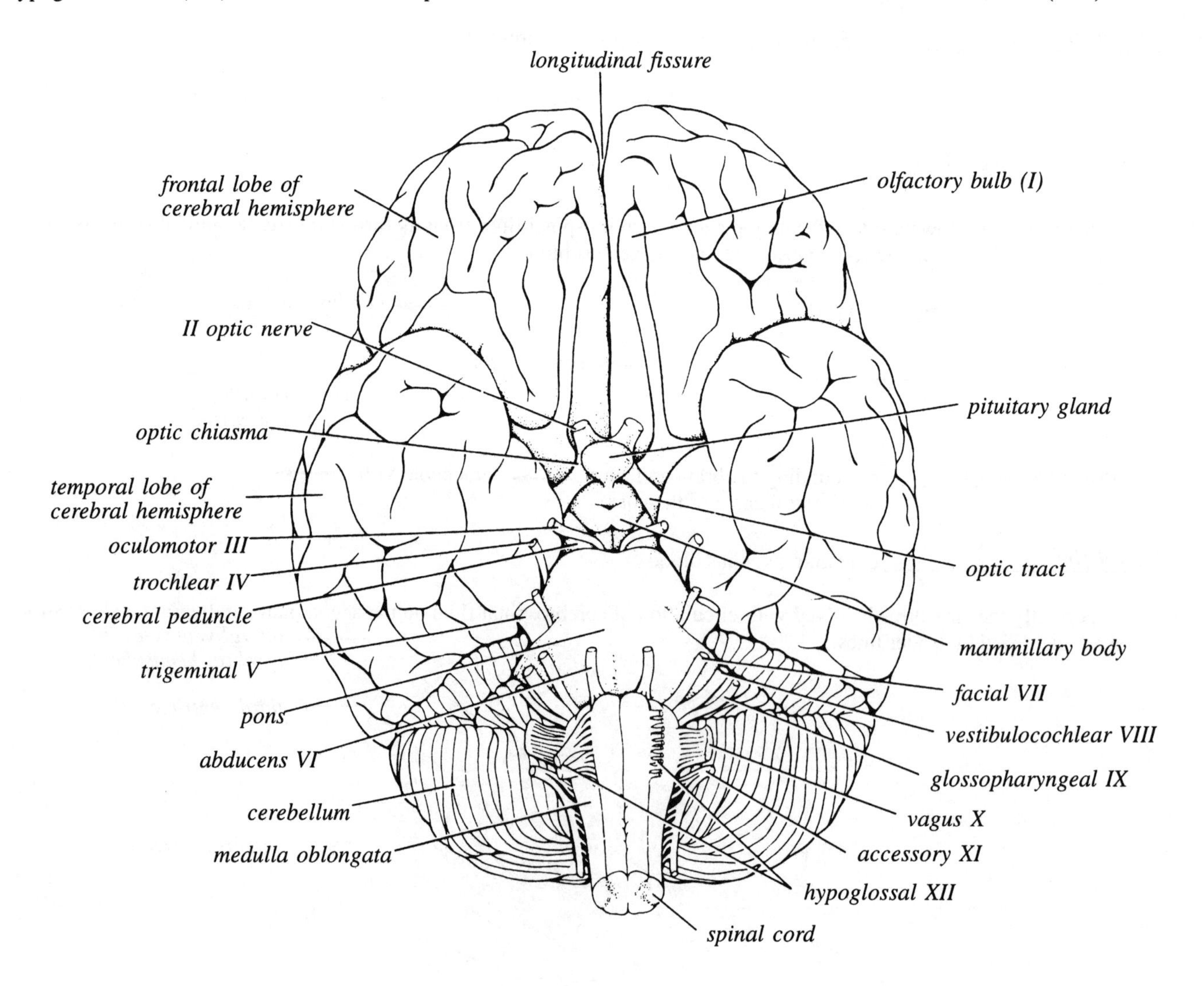

Dissection of Sheep Brain

12. In your own words, describe the relative hardness of the sheep brain tissue as noticed when you were cutting into it.

The sheep brain tissue is fairly soft.

Because formalin hardens all tissue, what conclusions might you draw about the relative hardness and texture of living brain tissue? *The human brain is a relatively soft tissue.*

13. How does the relative size of the cerebral hemispheres compare in sheep and human brains?

The cerebral hemispheres in human brains are larger.

What is the significance of this difference? *Humans have a more developed area for speech, language, conscious thought, interpretation, and other functions associated with the cerebrum.*

14. What is the significance of the fact that the olfactory bulbs are much larger in the sheep brain than in the human brain?

The sense of smell is more important as a protective and food-getting sense in sheep.

NAME ______________________ LAB TIME/DATE ______________________

REVIEW SHEET
exercise
15

Spinal Cord and Spinal Nerves

Anatomy of the Spinal Cord

1. Complete the following statements by inserting the proper anatomical terms in the answer blanks.

   The superior boundary of the spinal cord is at the level of the *foramen magnum*, and its inferior boundary is at the level of vertebra L_2. The collection of spinal nerves traveling in the vertebral canal below the terminus of the spinal cord is called the *cauda equina*.

2. Using the terms below, correctly identify on the diagram all structures provided with leader lines.

anterior (ventral) horn	dorsal root of spinal nerve	spinal nerve
arachnoid mater	dura mater	ventral ramus of spinal nerve
central canal	lateral horn	ventral root of spinal nerve
dorsal ramus of spinal nerve	pia mater	white matter
dorsal root ganglion	posterior (dorsal) horn	

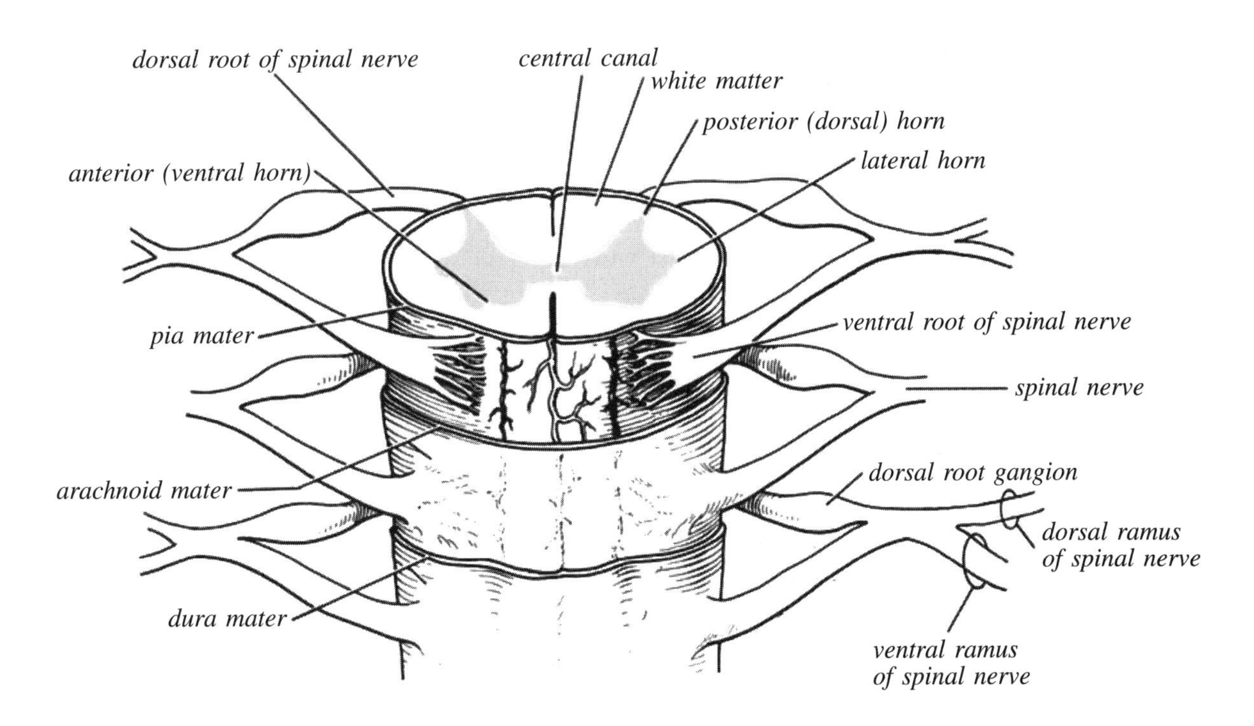

3. The spinal cord is enlarged in two regions, the *cervical* and the *lumbar* regions.

What is the significance of these enlargements? *This is where the nerves serving the upper and lower limbs leave the spinal cord.*

Spinal Nerves and Nerve Plexuses

4. In the human, there are 31 pairs of spinal nerves named according to the region of the vertebral column from which they issue. The spinal nerves are named below. Note, by number, the vertebral level at which they emerge:

cervical nerves $C_1 - C_8; T_1$

sacral nerves $L_4 - L_5; S_1 - S_4$

lumbar nerves $L_1 - L_4$

thoracic nerves $T_1 - T_{12}$

5. The ventral rami of spinal nerves C_1 through T_1 and T_{12} through S_4 form *ventral rami* which serve the *motor and sensory needs of the limbs* of the body. The ventral rami of T_2 through T_{12} run between the ribs toserve the *muscles of intercostal spaces and the skin and muscles of the anterior and lateral trunk*.

The dorsal rami of the spinal nerves serve *the skin and muscles of the posterior body trunk*.

6. What would happen (i.e., loss of sensory or motor function or both) if the following structures were damaged or transected?

1. dorsal root of a spinal nerve *sensory*
2. ventral root of a spinal nerve *motor*
3. anterior ramus of a spinal nerve *both*

7. Define *plexus:* *a complex nerve network that serves the motor and sensory needs*

8. Name the major nerves that serve the following body areas:

axillary 1. deltoid muscle

phrenic 2. diaphragm

sciatic 3. posterior thigh

common peroneal 4. lateral leg and foot

median 5. flexor muscles of forearm and some hand muscles

musculocutaneous 6. flexor muscles of arm

femoral 7. lower abdomen and anterior thigh

radial 8. triceps muscle

tibial 9. posterior leg and foot

NAME ______________________ LAB TIME/DATE ______________________

REVIEW SHEET
exercise
16

Human Reflex Physiology

The Reflex Arc

1. Define *reflex:* *automatic reaction to a stimulus, or a rapid, predictable, involuntary motor response to stimuli*

2. Name five essential components of a reflex arc: *receptor*, *sensory neuron*, *integration center*, *motor neuron*, and *effector*.

3. In general, what is the importance of reflex testing in a routine physical examination? *Reflexes determine the general health of the motor portion of the nervous system. Whenever reflexes are exaggerated, distorted or absent, nervous system disorders are indicated. Reflex changes often occur before the pathologic condition has become obvious in other ways.*

Somatic and Autonomic Reflexes

4. Use the key terms to complete the statements given below.

   Key:

Achilles reflex	gag reflex	plantar reflex
corneal reflex	patellar reflex	pupillary light reflex

   Reflexes classified as somatic reflexes include *Achilles*, *corneal*, *gag*, *patellar*, and *plantar*.

   Of these, the simple stretch reflexes are *patellar* and *Achilles*, and the superficial cord reflex is *plantar*.

   A reflex classified as an autonomic reflex is the *pupillary light reflex*.

5. Name two cord-mediated reflexes: *Achilles* and *plantar*

   Name two somatic reflexes in which the higher brain centers participate: *pupillary light reflex* and *corneal*

6. Trace the reflex arc, naming efferent and afferent nerves, receptors, effectors, and integration centers, for the following reflexes:

patellar reflex *Stretch receptors in the quadriceps muscle → afferent neuron → spinal cord → efferent neuron (femoral nerve L_2–L_4) → Quadriceps muscle of thigh*

Achilles reflex *stretch receptors in the Achilles tendon (calcaneal tendon) → afferent neuron → spinal cord → efferent neuron (tibial nerve S_1–S_2) → gastrocnemius muscle*

7. What was the effect of muscle fatigue on your ability to produce the patellar reflex?

The intensity of the response was less.

8. The pupillary light reflex and the corneal reflex illustrate the purposeful nature of reflex activity. Describe the protective aspect of each:

pupillary light reflex *protect the retina*

corneal reflex *protect the eyeball from damage*

9. Was the pupillary consensual response contralateral or ipsilateral? *Contralateral*

Why would such a response be of significant value in this particular reflex? *To see if cranial nerve III is intact*

10. Differentiate between the types of activities accomplished by somatic and autonomic reflexes. *Autonomic reflexes activate smooth muscles, cardiac muscle, and the glands of the body, and they regulate body functions, such as digestion and blood pressure. Somatic reflexes include all reflexes that stimulate skeletal muscles.*

NAME ______________________ LAB TIME/DATE ______________________

REVIEW SHEET
exercise
17

Special Senses

The Eye and Vision: Anatomy

1. Several accessory eye structures contribute to the formation of tears and/or aid in lubrication of the eyeball. Match the described accessory structures with their secretion by choosing answers from the key.

   Key: conjunctiva lacrimal glands tarsal glands

   conjunctiva 1. mucus

   tarsal glands 2. oil

   lacrimal glands 3. salt solution

2. The eyeball is wrapped in adipose tissue within the orbit. What is the function of the adipose tissue?

   Protection

3. Why may it be necessary to blow one's nose after having a good cry? *Tears released by lacrimal glands flush across the eyeball into the lacrimal canals to the lacrimal sacs to the nasolacrimal duct into the nasal cavity.*

4. What is a sty? *A sty is the inflammation of ciliary glands (modified sweat glands) that lie between the eyelashes and help lubricate the eyeball.*

   Conjunctivitis? *Inflammation of the conjunctiva, which is the mucus membrane that lines the internal surface of the eyelids and eyeball.*

5. What seven bones form the bony orbit? (Think! If you can't remember, check a skull or your textbook.)

   maxilla *palatine* *zygomatic*

   ethmoid *sphenoid* *frontal*

   lacrimal

6. Identify the lettered structures on the diagram by matching each letter with one of the terms to the right.

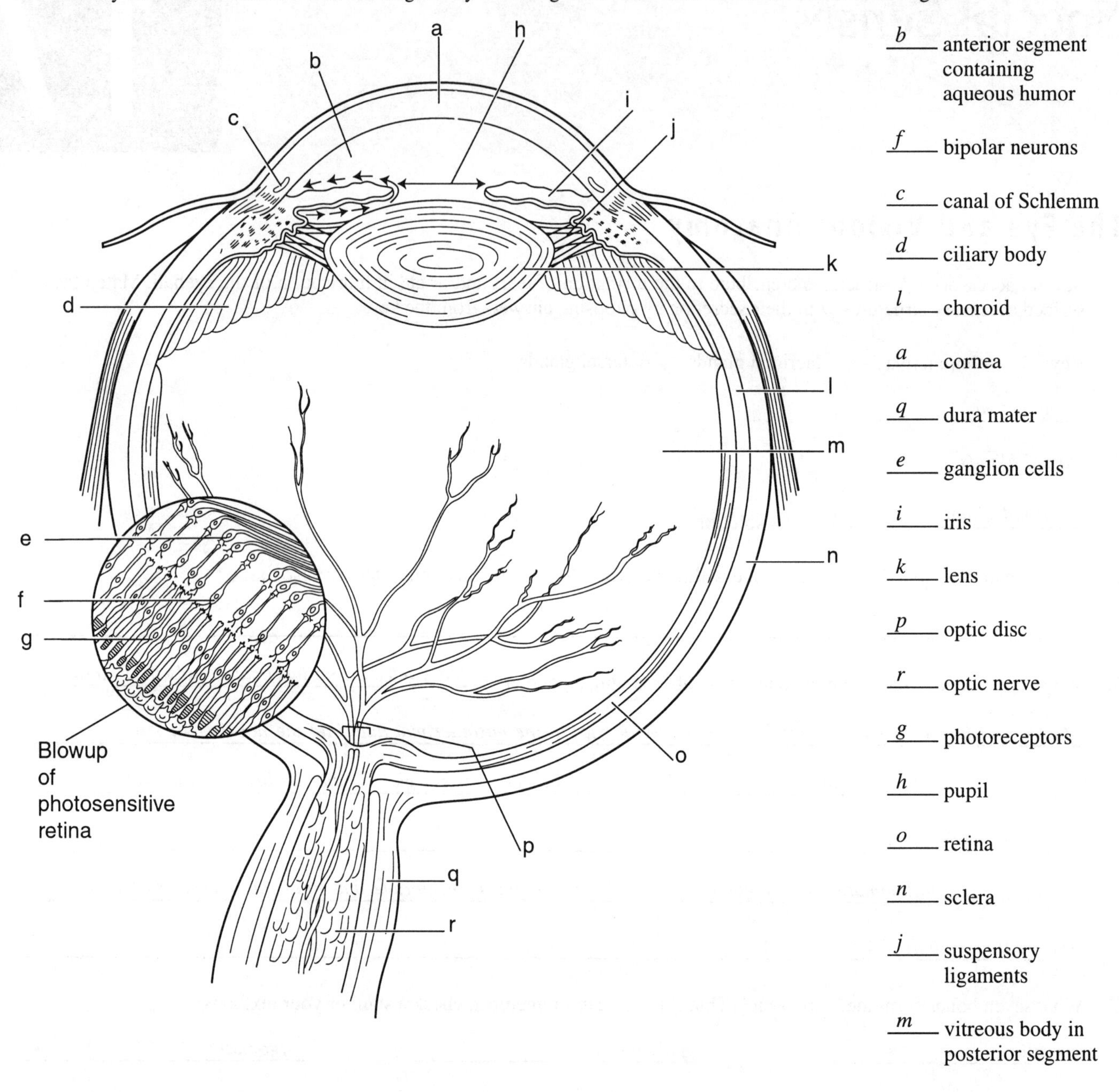

b anterior segment containing aqueous humor

f bipolar neurons

c canal of Schlemm

d ciliary body

l choroid

a cornea

q dura mater

e ganglion cells

i iris

k lens

p optic disc

r optic nerve

g photoreceptors

h pupil

o retina

n sclera

j suspensory ligaments

m vitreous body in posterior segment

Notice the arrows drawn close to the left side of the iris in the diagram above. What do they indicate?

Pupillary distance

7. Match the key responses with the descriptive statements that follow.

Key:	aqueous humor	cornea	lens	sclera
	canal of Schlemm	fovea centralis	optic disc	suspensory ligament
	choroid	iris	retina	vitreous humor
	ciliary body			

suspensory ligament 1. attaches the lens to the ciliary body *aqueous humor* 2. fluid filling the anterior segment of the eye

optic disc 3. the blind spot

iris 4. contains muscle that controls the size of the pupil

canal of Schlemm 5. drains the aqueous humor from the eye

retina 6. "sensory" tunic

vitreous humor 7. substance occupying the posterior segment of the eyeball

choroid 8. forms most of the pigmented vascular tunic

fovea centralis 9. tiny pit in the macula lutea; contains only cones

lens 10. important light-bending structure of the eye; shape can be modified

cornea 11. anterior transparent part of the fibrous tunic

sclera 12. composed of tough, white, opaque, fibrous connective tissue

8. The intrinsic eye muscles are under the control of which of the following? (Circle the correct response.)

(autonomic nervous system) somatic nervous system

Dissection of the Cow (Sheep) Eye

9. What modification of the choroid that is not present in humans is found in the cow eye? *Choroid appears irridescent due to special reflecting surface (tapetum lucidum).*

What is its function? *Reflects light within the eye in animals that live under conditions of low-intensity light*

10. Describe the appearance of the retina. *Appears as a delicate white, crumpled membrane that separated easily from the pigmented choroid*

At what point is it attached to the posterior aspect of the eyeball? *At the optic disc*

Visual Tests and Experiments

11. Use terms from the key to complete the statements concerning near and distance vision.

Key: contracted decreased increased relaxed taut

During distance vision: The ciliary muscle is *relaxed*, the suspensory ligament is *taut*, the convexity of the lens is *decreased*, and light refraction is *decreased*. During close vision: The ciliary muscle is *contracted*, the suspensory ligament is *relaxed*, lens convexity is *increased*, and light refraction is *increased*.

12. Explain why the part of the image hitting the blind spot is not seen. *There are no photoreceptors at the blind spot.*

13. Match the terms in column B with the descriptions in column A:

	Column A	Column B
refraction	1. light bending	accommodation
accommodation	2. ability to focus for close (under 20 ft) vision	astigmatism
emmetropia	3. normal vision	convergence
hyperopia	4. inability to focus well on close objects (farsightedness)	emmetropia
myopia	5. nearsightedness	hyperopia
astigmatism	6. blurred vision due to unequal curvatures of the lens or cornea	myopia
convergence	7. medial movement of the eyes during focusing on close objects	refraction

14. Record your Snellen eye test results below:

Left eye (without glasses) ______________ (with glasses) ______________

Right eye (without glasses) ______________ (with glasses) ______________

Is your visual acuity normal, less than normal, or better than normal? ______________

Explain. ______________

Explain why each eye is tested separately when using the Snellen eye chart. *To check the visual acuity of each eye*

Explain 20/40 vision. *I can see at 20′ what is normally seen at 40′.*

Explain 20/10 vision. *I can see at 20′ what is normally seen at 10′.*

15. Define *astigmatism:* *Blurred vision due to irregularities in the curvature of the lens and/or cornea*

16. Record the distance of your near point of accommodation as tested in the laboratory:

right eye ______________ left eye ______________

Is your near point within the normal range for your age? *(10 cm in young adults)*

17. How can you explain the fact that we see a great range of colors even though only three cone types exist?

Interpretation of the intermediate colors of the visible light spectrum is a result of simultaneous input from more than one cone type.

18. In the experiment on the convergence reflex, what happened to the position of the eyeballs as the object was moved closer to the subject's eyes? *Eyeballs move medially*

What extrinsic eye muscles control the movement of the eyes during this reflex? *Medial rectus*

What is the value of this reflex? *These muscles control eye movement and make it possible to keep objects focused on the fovea centralis as they move closer, which is essential for near vision.*

What would be the visual result of an inability of these muscles to function?

The inability to focus on near objects

19. Many college students struggling through mountainous reading assignments are told that they need glasses for "eyestrain." Why is it more of a strain on the extrinsic and intrinsic eye muscles to look at close objects than at far objects?

To make close vision possible, ciliary muscles must contract and both eyeballs must be kept looking medially and inferiorly.

The Ear and Hearing and Balance: Anatomy

20. Select the terms from column B that apply to the column A descriptions. Some terms are used more than once.

Answer	Column A
anvil (incus), *hammer (malleus)*, *stirrup (stapes)*	1. collectively called the ossicles
vestibule, *semicircular canals*	2. ear structures involved with balance
tympanic membrane	3. transmits sound vibrations to the ossicles
semicircular canals	4. three circular passages, each in a different plane of space
oval window	5. transmits the vibratory motion of the stirrup to the fluid in the inner ear
auditory (pharyngotympanic) tube	6. passage between the throat and the tympanic cavity
endolymph	7. fluid contained within the membranous labyrinth

Column B

auditory (pharyngotympanic) tube

anvil (incus)

cochlea

endolymph

external acoustic meatus

hammer (malleus)

oval window

perilymph

pinna

round window

semicircular canals

stirrup (stapes)

tympanic membrane

vestibule

21. Identify all indicated structures and ear regions that are provided with leader lines or brackets in the following diagram.

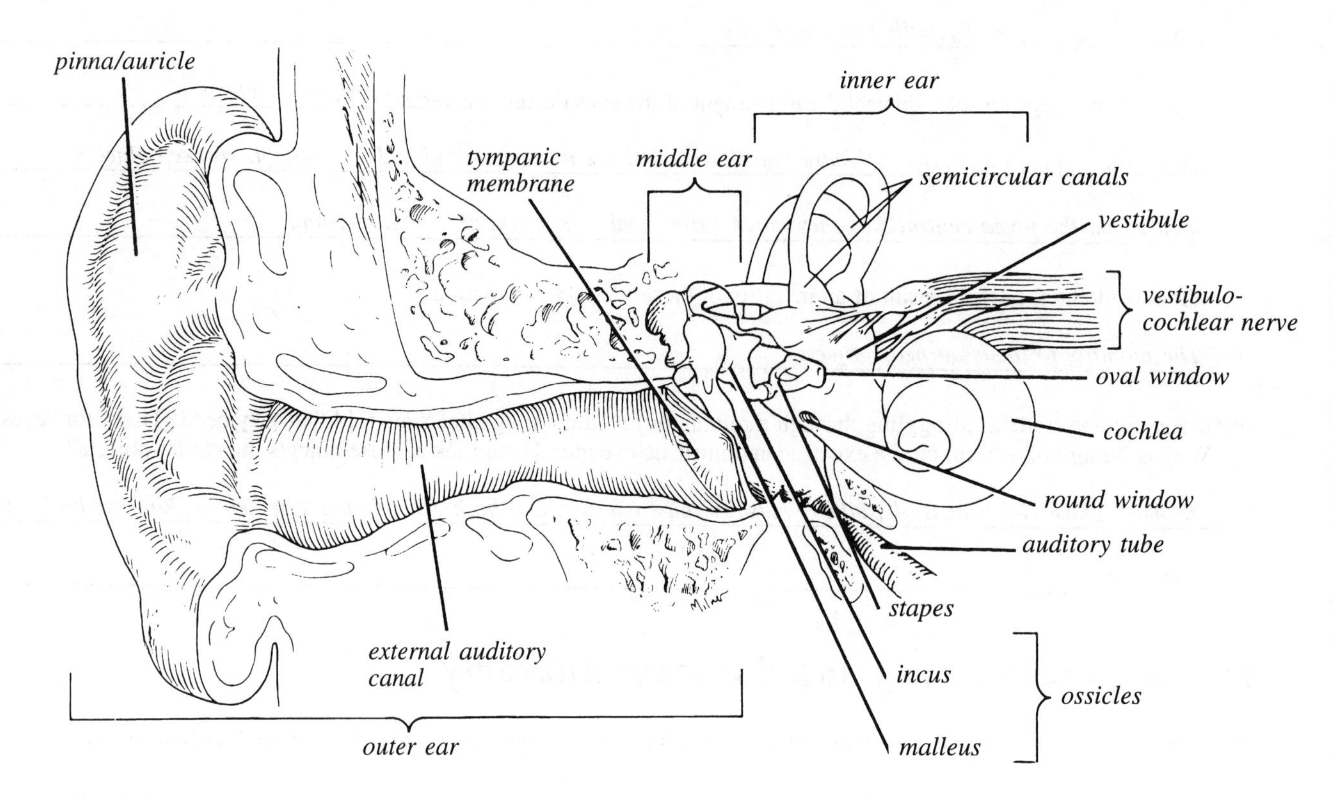

22. Match the membranous labyrinth structures listed in column B with the descriptive statements in column A

Answer	Column A	Column B
cochlear duct	1. contains the organ of Corti	ampulla
vestibular sacs, *otolith*	2. sites of the maculae	basilar membrane
basilar membrane	3. hair cells of organ of Corti rest on this membrane	cochlear duct
tectorial membrane	4. gel-like membrane overlying the hair cells of the organ of Corti	cochlear nerve
ampulla	5. contains the cristae ampullaris	cupula
otoliths, *vestibular nerve*, *vestibular sacs*	6. function in static equilibrium	otoliths
		semicircular ducts
semicircular ducts, *ampulla*, *cupula*, *vestibular nerve*	7. function in dynamic equilibrium	tectorial membrane
		vestibular nerve
cochlear nerve	8. carries auditory information to the brain	vestibular sacs
cupula	9. gelatinous cap overlying hair cells of the crista ampullaris	
otoliths	10. grains of calcium carbonate in the maculae	

23. Describe how sounds of different frequency (pitch) are differentiated in the cochlea. *In general, high-pitch sounds disturb receptor cells close to the oval window, whereas low-pitch sounds stimulate specific hair cells further along the cochlea.*

24. Explain the role of the endolymph of the semicircular canals in activating the receptors during angular motion.

During angular motion, the endolymph lags behind and moves in the opposite direction, pushing the cupula in a direction opposite to the body's motion. This stimulates the hair cells, and impulses are transmitted up the vestibular nerve to the cerebellum.

25. Explain the role of the otoliths in perception of static equilibrium (head position). __________

As the head moves, the otoliths roll in response to changes in the pull of gravity. This movement creates a pull on the gel, which in turn bends the hair cells, sending impulses along the vestibular nerve to the cerebellum informing it of the head position.

Hearing and Balance Tests

26. Was the auditory acuity measurement made during the experiment on page 148 the same or different for both ears?

__________ What factors might account for a difference in the acuity of the two ears?

Several factors may be inflammation or infection in the ear drum, ear wax buildup, unequal air pressure, damaged hair cell receptors.

27. During the sound localization experiment on page 148, in which position(s) was the sound least easily located?

The positions include above the head, in front and in back of head.

How can this observation be explained? *The ability to localize sound source depends on difference in loudness and time of arrival of sound at each ear. From these positions, the loudness and arrival time are equal.*

28. When the tuning fork handle was pressed to your forehead during the Weber test, where did the sound seem to originate?

From the ears

Where did it seem to originate when one ear was plugged with cotton? *From the side plugged with cotton*

How do sound waves reach the cochlea when conduction deafness is present? __________

By bone conduction

29. The Rinne test evaluates an individuals's ability to hear sounds conducted by air or bone. Which is typical of normal hearing? *Air conduction is the normal conduction route.*

30. Define *nystagmus:* *Involuntary rolling of the eyes in any direction or the trailing of the eyes slowly in one direction, followed by their rapid movement in the opposite direction*

31. What is the usual reason for conducting the Romberg test? *It determines the soundness of the dorsal white column of the spinal cord, which transmits impulses to the brain from the proprioceptors involved with posture.*

Was the degree of sway greater with the eyes open or closed? *Closed*

Why? *Sight enhances the information provided by the semicircular canals.*

32. Normal balance, or equilibrium, depends on input from a number of sensory receptors. Name them.

Hair cells in semicircular canal, proprioceptors, sight (retina)

Chemical Senses: Localization and Anatomy of Olfactory and Taste Receptors

33. Describe the cellular makeup and the location of the olfactory epithelium. *It occupies an area about 2.5 cm in the roof of each nasal cavity with bipolar neurons and support epithelial cells.*

34. Name three sites where receptors for taste are found, and circle the predominant site:

tongue (circled), *soft palate and pharynx*, and

inner surface of the checks

35. Describe the cellular makeup and arrangement of a taste bud. (Use a diagram, if helpful.)

See Figure 17.15(c)

Taste and Smell Experiments

36. Taste and smell receptors are both classified as *chemoreceptors* because they both respond to *chemicals in solution*

37. Why is it impossisble to taste substances with a dry tongue? *The chemicals must be in solution to stimulate the chemoreceptors.*

38. Name three factors that influence our appreciation of foods. Substantiate each choice with an example from the laboratory experience.

Smell Substantiation *Peppermint, wintergreen, and clove oil must be smelled to distinguish the flavor.*

Texture Substantiation *The foods with the same texture could not be distinguished by texture alone.*

Temperature Substantiation *The colder the tongue, the harder it is to distinguish tastes.*

Expand on your explanation and choices by explaining why a cold, greasy hamburger is unappetizing to most people.

The smell, texture, and temperature are unappealing.

39. How palatable is food when you have a cold? *Less than normal*

Explain. *With a cold, your sense of smell is decreased.*

NAME ______________________ LAB TIME/DATE ______________________

REVIEW SHEET
exercise
18

Functional Anatomy of the Endocrine Glands

Gross Anatomy and Basic Function of the Endocrine Glands

1. The endocrine and nervous systems are major regulating systems of the body. However, the nervous system has been compared to an airmail delivery system and the endocrine system to the pony express. Briefly explain this comparison.

   The nervous system uses nerve impulses to bring rapid control. The endocrine system acts much more slowly. The endocrine glands produce hormones and release them into the blood to travel to relatively distant target organs.

2. Define *hormone:* *Chemical messengers secreted by endocrine glands; responsible for specific regulatory effects on certain parts or organs*

3. Chemically, hormones belong chiefly to two molecular groups, the *amino acids* and the *steroids*.

4. Define *target organ:* *An organ upon which a specific hormone affects*

5. Why don't all tissues respond to all hormones? *Specific protein receptors must be present on the plasma membrane or within the cells for the tissue cells to respond*

6. Identify the endocrine organ described by the following statements:

   thyroid 1. located in the throat; bilobed gland connected by an isthmus

   adrenal 2. found close to the kidney

   pancreas 3. a mixed gland, located close to the stomach and small intestine

   testes 4. paired glands suspended in the scrotum

   parathyroid 5. ride "horseback" on the thyroid gland

   ovaries 6. found in the pelvic cavity of the female, concerned with ova and female hormone production

   thymus 7. found in the upper thorax overlying the heart; large during youth

   pineal 8. found in the roof of the third ventricle

7. Although the pituitary gland is often referred to as the master gland of the body, the hypothalamus exerts some control over the pituitary gland. How does the hypothalamus control both anterior and posterior pituitary functioning?

The anterior pituitary is controlled by releasing or inhibiting hormones produced by the hypothalamus. The posterior pituitary stores two hormones transported to it along axons from the hypothalamus.

8. For each statement describing hormonal effects, identify the hormone(s) involved by choosing a number from key A, and note the hormone's site of production with a letter from key B. More than one hormone may be involved in some cases.

Key A:

1. ACTH
2. ADH
3. aldosterone
4. calcitonin
5. cortisone
6. epinephrine
7. estrogens
8. FSH
9. glucagon
10. GH
11. insulin
12. LH
13. melatonin
14. MSH
15. oxytocin
16. progesterone
17. prolactin
18. PTH
19. serotonin
20. testosterone
21. thymosin
22. T_4/T_3
23. TSH

Key B:

a. adrenal cortex
b. adrenal medulla
c. anterior pituitary
d. hypothalamus
e. ovaries
f. pancreas
g. parathyroid glands
h. pineal gland
i. posterior pituitary
j. testes
k. thymus gland
l. thyroid gland

22, *l* 1. basal metabolism hormone

21, *k* 2. helps program the immune system

4, *l* and *18*, *g* 3. regulate blood calcium levels

6, *b* and *5*, *a* 4. released in response to stressors

7, *e* and *20*, *j* 5. drives development of secondary sexual characteristics

8, *c*; *12*, *c*; *1*, *c*; and *23*, *c* 6. regulate the function of another endocrine gland

6, *b.* 7. mimics the sympathetic nervous system

11, *f* and *9*, *f* 8. regulate blood glucose levels; produced by the same "mixed" gland

7, *e* and *16*, *e* 9. directly responsible for regulating the menstrual cycle

2, *i* and *3*, *a* 10. helps maintain salt and water balance in the body fluids

15, *i* 11. involved in milk ejection

9. Name the hormone whose production in *inadequate* amounts results in the following conditions. (Use your textbook as necessary.)

testosterone/estrogen 1. poor development of secondary sex characteristics

parathyroid hormone (PTH) 2. tetany

insulin ______ 3. diabetes mellitus

growth hormone (GH) ______ 4. abnormally small stature, normal proportions

thyroxine (T_4) ______ 5. myxedema

Observing the Effects of Hyperinsulinism

10. Briefly explain what was happening within the fish's system when the fish was immersed in the insulin solution.

The fish absorbed the insulin, putting the fish into insulin shock, where blood glucose is transported into the cells leaving an extremely low amount of glucose in the bloodstream.

11. What is the mechanism of the recovery process you observed? *Negative feedback mechanism*

12. What would you do to help a friend who had inadvertently taken an overdose of insulin? *Give them candy or sugared drinks.* Why? *So that the excess insulin will be used to increase the ability to transport glucose into cells*

13. What is a glucose tolerance test? (Use an appropriate reference, as necessary, to answer this question.)

Glucose tolerance test determines how well your body regulates blood glucose levels by ingesting a large amount of glucose and periodically (1–3 hrs) taking blood and urine samples and testing the amount of glucose found in the samples.

NAME ______ LAB TIME/DATE ______

REVIEW SHEET
exercise
19

Blood

Composition of Blood

1. What is the blood volume of an average-size adult? *5.5* liters

2. What determines whether blood is bright red or a dull brick-red?

Depends on the amount of oxygen the red-blood cell is carrying

3. Use the key to identify the cell type(s) or blood elements that fit the following descriptive statements.

Key:

basophil	formed elements	monocyte
eosinophil	lymphocyte	neutrophil
erythrocyte	megakaryocyte	plasma

neutrophil 1. its name means "neutral-loving," a phagocyte

basophil, *eosinophil*, and *neutrophil* 2. granulocytes

erythrocyte 3. also called a red blood cell

monocyte, *lymphocyte* 4. agranulocytes

megakaryocyte 5. ancestral cell of platelets

formed elements 6. basophils, eosinophils, erythrocytes, lymphocytes, megakaryocytes, monocytes, and neutrophils are all examples of these

eosinophil 7. number rises during parasite infections

basophil 8. releases a vasodilator; the least abundant WBC

erythrocyte 9. transports oxygen

plasma 10. primarily water, noncellular; the fluid matrix of blood

monocyte 11. phagocyte in chronic infections

basophil, *eosinophil*, *lymphocyte*,

monocyte, *neutrophil* 12. also called white blood cells

4. List three classes of nutrients normally found in plasma: *Glucose, fatty acids*,

vitamins, and *amino acids*

Name two gases: O_2 and CO_2

Name three ions: *Sodium*, *potassium*, and *calcium*

5. Describe the consistency and color of the plasma you observed in the laboratory. ______

Viscous; slippery; yellow-straw

6. What is the average life span of a red blood cell? How does its anucleate condition affect this life span?

Life span is 100–120 days. Because they are anucleate, they are unable to reproduce.

7. From memory, describe the structural characteristics of each of the following blood cell types as accurately as possible, and note the percentage of each in the total white blood cell population.

eosinophils *1–4% Nucleus: bilobed (figure 8)*

Cytoplasm: large red-orange granules

neutrophils *40–70% Nucleus: 3–7 lobes*

Cytoplasm: pale lilac w/ fine red/blue granules

lymphocytes *20–45% Nucleus: dark blue to purple, spherical*

Cytoplasm: sparse; thin blue rim around nucleus

basophils *<1% Nucleus: U or S shaped*

Cytoplasm: deep purple granules

monocytes *4–8% Nucleus: dark blue, kidney shaped*

Cytoplasm: abundant gray-blue

8. Correctly identify the blood pathologies described in column A by matching them with selections from column B:

	Column A	Column B
leukocytosis	1. abnormal increase in the number of WBCs	anemia
polycythemia	2. abnormal increase in the number of RBCs	leukocytosis
anemia	3. condition of too few RBCs or of RBCs with hemoglobin deficiencies	leukopenia
		polycythemia
leukopenia	4. abnormal decrease in the number of WBCs	

Hematologic Tests

9. Broadly speaking, why are hematologic studies of blood so important in the diagnosis of disease?

These tests are useful diagnostic tools for the physician because blood composition reflects the status of many body functions and malfunctions.

10. In the chart that follows, record information from the blood tests you conducted. Complete the chart by recording values for healthy male adults and indicating the significance of high or low values for each test.

Test	Student test results	Normal values (healthy male adults)	Significance High values	Low values
Total WBC count	—	*4,000–11,000/mm^3*	*Leukocytosis*	*Leukopenia*
Total RBC count	—	*4–6 million/mm^3*	*Polycythemia*	*Anemia*
Hematocrit		*47.0% ± 7%*	*Polycythemia*	*Anemia*
Hemoglobin determination		*13–18 g/100 mL*	*Polycythemia*	*Anemia*
Coagulation time		*3–6 minutes*	*Decreased clotting factors*	*Increased clotting factors*

11. Define *hematocrit:* *The percentage of erythrocytes to total blood volume*

12. If you had a high hematocrit, would you expect your hemoglobin determination to be high or low?

High Why? *The higher percentage of erythrocytes would normally mean a higher amount of hemoglobin, which is the protein inside the erythrocyte responsible for oxygen transport.*

13. If your blood clumped with both anti-A and anti-B sera, your ABO blood type would be *AB*. To what ABO blood groups could you give blood? *AB* From which ABO donor types could you receive blood? *All other types*

Which ABO blood type is most common? *O* Least common? *AB*

14. Explain why an Rh-negative person does not have a tranfusion reaction on the first exposure to Rh-positive blood but *does* have a reaction on the second exposure. *Rh-negative persons who receive transfusions of Rh-positive blood become sensitized by the Rh antigens of the donor RBC's, and their systems begin to produce anti-Rh antibodies. On later exposures to Rh-positive blood, typical transfusion reactions occur.*

What happens when an ABO blood type is mismatched for the first time? *The plasma proteins (antibodies) will bind to RBC's bearing different antigens, causing them to be clumped, agglutinated, and eventually hemolyzed.*

15. Assume the blood of two patients has been typed for ABO blood type.

Typing results
Mr. Adams:

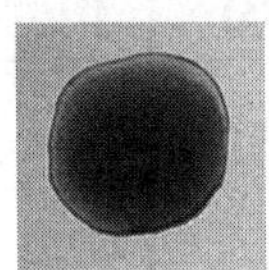

Blood drop
and anti-A serum

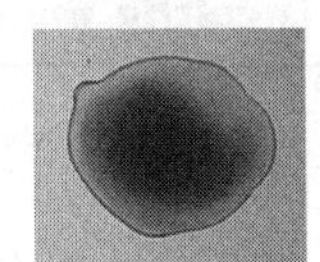

Blood drop
and anti-B serum

Typing results
Mr. Calhoon:

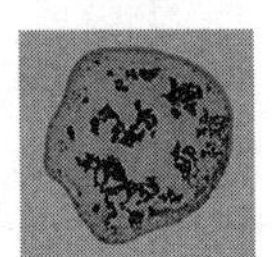

Blood drop
and anti-A serum

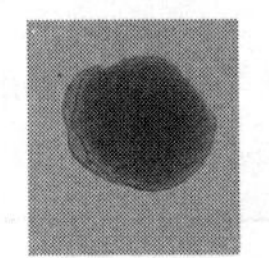

Blood drop
and anti-B serum

On the basis of these results, Mr. Adams has type ___0___ blood, and Mr. Calhoon has type ___A___ blood.

NAME ______________________ LAB TIME/DATE ______________________

REVIEW SHEET
exercise
20

Anatomy of the Heart

Gross Anatomy of the Human Heart

1. An anterior view of the heart is shown here. Identify each numbered structure by writing its name on the correspondingly numbered line:

1. *Right atrium*
2. *Left atrium*
3. *Right ventricle*
4. *Left ventricle*
5. *Superior vena cava*
6. *Inferior vena cava*
7. *Aorta*
8. *Brachiocephalic artery*
9. *Left common carotid artery*
10. *Left subclavian artery*
11. *Pulmonary trunk*
12. *Right pulmonary artery*
13. *Left pulmonary artery*
14. *Ligamentum arteriosum*
15. *Right pulmonary veins*
16. *Left pulmonary veins*
17. *Right atrioventricular groove*
18. *Right coronary artery*
19. *Left atrioventricular groove*
20. *Left coronary artery*
21. *Apex*

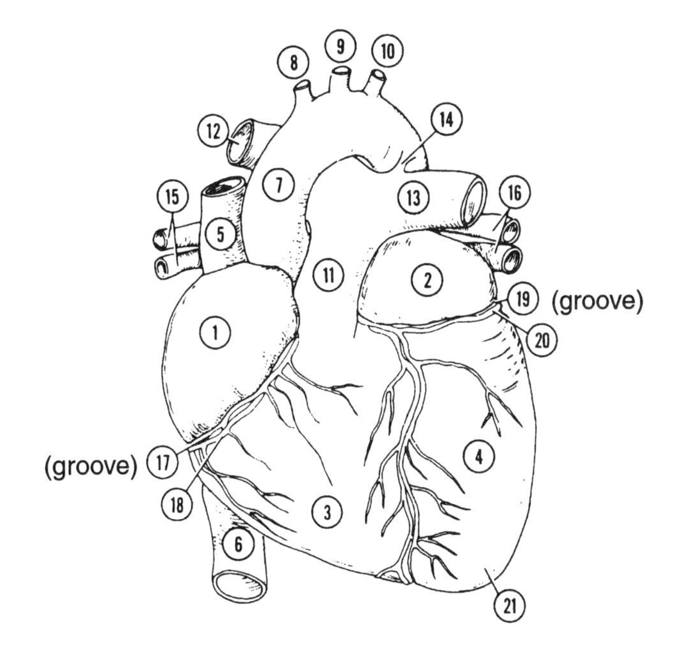

2. What is the function of the fluid that fills the pericardial sac? *Serous fluid allows the heart to beat in a relatively frictionless environment.*

3. Match the terms in the key to the descriptions provided below.

Answer	Description	Key:
coronary sinus	1. drains blood into the right atrium	atria
atria	2. superior heart chambers	coronary arteries
ventricles	3. inferior heart chambers	coronary sinus
epicardium	4. visceral pericardium	endocardium
atria	5. "anterooms" of the heart	epicardium
myocardium	6. equals cardiac muscle	myocardium
coronary arteries	7. provide nutrient blood to the heart muscle	ventricles
endocardium	8. lining of the heart chambers	
ventricles	9. actual "pumps" of the heart	

4. What is the function of the valves found in the heart? *To inhibit the backflow of blood; provide one-way blood flow through heart chambers*

5. Can the heart function with leaky valves? (Think! Can a water pump function with leaky valves?) *Poorly*

6. What is the role of the chordae tendineae? *Anchors the cusps of the valve to the ventricular walls*

Pulmonary, Systemic, and Cardiac Circulations

7. A simple schematic of a so-called general circulation is shown below. What part of the circulation is missing from this diagram?

Pulmonary circulation

Add to the diagram as best you can to make it depict a complete systemic/pulmonary circulation and re-identify "general circulation" as the correct subcirculation.

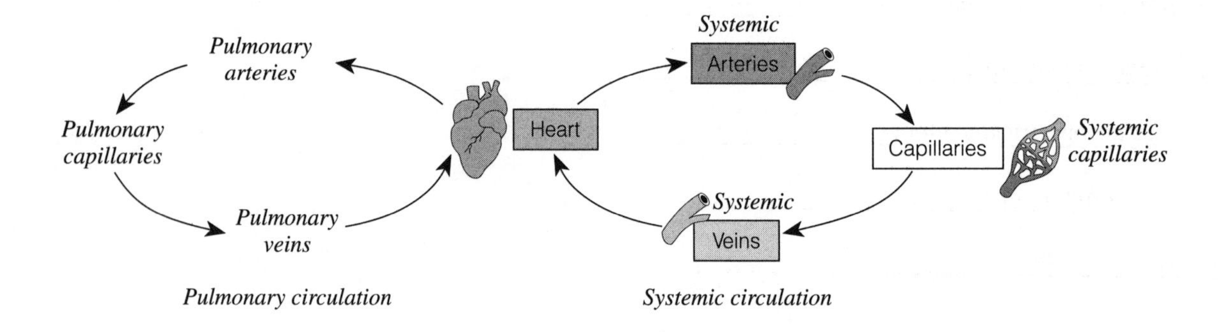

8. Differentiate clearly between the roles of the pulmonary and systemic circulations. *The pulmonary circulation sends CO_2-rich blood to the lung to unload CO_2 and pick up O_2 to carry back to the heart. Systemic circulation takes O_2-rich blood from the heart through the body tissues and back to the heart.*

9. Complete the following scheme of circulation in the human body:

Right atrium through the tricuspid valve to the *right ventricle* through the *pulmonary semilunar* valve to the pulmonary trunk to the *pulmonary arteries* to the capillary beds of the lungs to the *pulmonary veins* to the *left atrium* of the heart through the *mitral (bicuspid)* valve to the *left ventricle* through the *aortic semilunar* valve to the *aorta* to the systemic arteries to the *capillaries* of the tissues to the systemic veins to the *inferior vena cava* and *superior vena cava* entering the right atrium of the heart.

10. If the mitral valve does not close properly, which circulation is affected? *Systemic*

11. Why might a thrombus (blood clot) in the anterior descending branch of the left coronary artery cause sudden death? *A thrombus would block oxygen-rich blood from being supplied to the anterior ventricular walls and the laterodorsal part of the left side of the heart.*

Microscopic Anatomy of Cardiac Muscle

12. Add the following terms to the photo of cardiac muscle at the right:

intercalated discs

nucleus of cardiac fiber

striations

cardiac muscle fiber

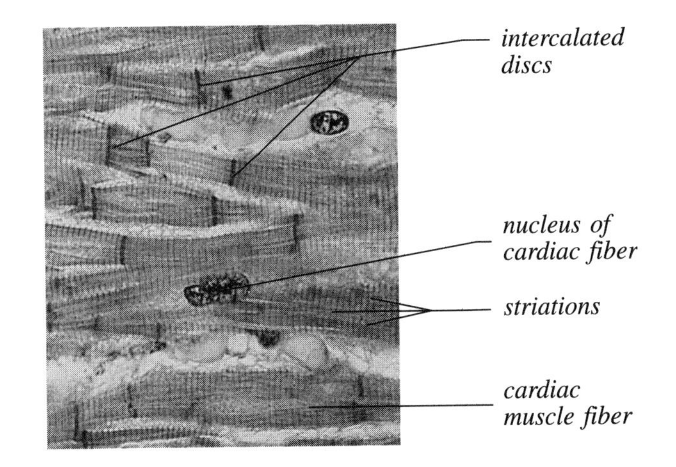

13. What role does the unique structure of cardiac muscle play in its function? (**Note:** before attempting a response, *describe* the unique anatomy.) *The cardiac cells are arranged in spiral or figure-8 shaped bundles. When the heart contracts, blood is forced into the arteries leaving the heart. The cardiac cells are branching and interdigitate. These two structural features provide a continuity to cardiac muscle allowing coordinated heart activity.*

Dissection of the Sheep Heart

14. During the sheep heart dissection, you were asked initially to identify the right and left ventricles without cutting into the heart. During this procedure, what differences did you observe between the two chambers?

The left ventricle is thicker and more solid. The right ventricle is much thinner and feels somewhat flabby when compressed.

How would you say this structural difference reflects the relative functions of these two heart chambers?

This difference reflects the greater demand placed on the left ventricle, which must pump blood through the much longer systemic circulation.

15. Semilunar valves prevent backflow into the *ventricles*; AV valves prevent backflow into the *atrium*. Using your own observations, explain how the operation of the semilunar valves differs from that of the AV valves. *The semilunar valves do not have chordae tendineae and the three symmetrical cusps are relatively stable. When each semilunar valve cusp closes, it supports the other two cusps.*

16. Two remnants of fetal structures are observable in the heart—the ligamentum arteriosum and the fossa ovalis. What were they called in the fetal heart, where was each located, and what common purpose did they serve as functioning fetal structures?

ligamentum arteriosum: *The ligamentum arteriosum was called the ductus arteriosus and located between the pulmonary trunk and the aorta.*

fossa ovalis: *The fossa ovalis was called the foramen ovale and located between the right and left atrium. Since the fetal lungs are nonfunctional, both shunts allowed the blood to entirely bypass the lungs.*

NAME ______________________ LAB TIME/DATE ______________________

REVIEW SHEET
exercise
21

Anatomy of Blood Vessels

Microscopic Structure of the Blood Vessels

1. Use the key choices to identify the blood vessel tunic described.

Key: tunica intima tunica media tunica externa

tunic intima 1. most internal tunic

tunic media 2. bulky middle tunic contains smooth muscle and elastin

tunic intima 3. its smooth surface decreases friction

tunic intima 4. tunic of capillaries

tunic intima, *tunic media*, *tunic externa* 5. tunic(s) of arteries and veins

tunic media 6. tunic that is especially thick in arteries

tunic externa 7. most superficial tunic

2. Servicing the capillaries is the basic function of the organs of the circulatory system. Explain this statement.

Only the capillaries directly serve the needs of the body's cells. It is through the capillary walls that exchanges are made between tissue cells and blood. Respiratory gases, nutrients, and wastes move along diffusion gradients.

3. Cross-sectional views of an artery and of a vein are shown here. Identify each. Also respond to the related questions that follow.

artery *vein*

Which of these vessels may have valves? *Veins*

Which of these vessels depends on its elasticity to propel blood along? *Arteries*

Which depends on the respiratory and muscular pumps? *Veins* Explain this dependence.

As skeletal muscles surrounding the veins contract and relax, the blood is milked through the veins toward the heart. Pressure changes that occur in the thorax during breathing also aid blood return.

4. Why are the walls of arteries relatively thicker than those of the corresponding veins? *Because arteries are closer to the pumping action of the heart, they must be able to expand as blood is propelled into them and then recoil passively as the blood flows off into the circulation during diastole. Arterial walls must be strong and resilient to withstand blood pressure fluctuations.*

Major Systemic Arteries and Veins of the Body

5. Use the key on the right to identify the arteries or veins described on the left.

brachiocephalic 1. the arterial system has one of these; the venous system has two

coronary 2. these arteries supply the myocardium

internal carotid 3. the more anterior artery pair serving the brain

great saphenous 4. longest vein in the body

dorsalis pedis 5. artery on the foot checked after leg surgery

deep femoral 6. serves the posterior thigh

phrenic 7. supplies the diaphragm

brachial 8. formed by the union of the radial and ulnar veins

basilic, *cephalic* 9. two superficial veins of the arm

renal 10. artery serving the kidney

gonadal 11. testicular or ovarian veins

inferior mesenteric 12. artery that supplies the distal half of the large intestine

common iliac 13. drains the pelvic organs and lower limbs

common iliac 14. what the external iliac vein drains into in the pelvis

brachial 15. major artery serving the arm

superior mesenteric 16. supplies most of the small intestine

popliteal 17. what the femoral artery becomes at the knee

celiac trunk 18. an arterial trunk that has three major branches, which run to the liver, spleen, and stomach

external carotid 19. major artery serving the skin and scalp of the head

anterior tibial, *posterior tibial* 20. two veins that join, forming the popliteal vein

radial 21. artery generally used to take the pulse at the wrist

Key:
anterior tibial
basilic
brachial
brachiocephalic
celiac trunk
cephalic
common carotid
common iliac
coronary
deep femoral
dorsalis pedis
external carotid
femoral
gonadal
great saphenous
inferior mesenteric
internal carotid
internal iliac
fibularis
phrenic
popliteal
posterior tibial
radial
renal
subclavian
superior mesenteric
vertebral

6. The human arterial and venous systems are diagrammed on this page and the next. Identify all indicated blood vessels.

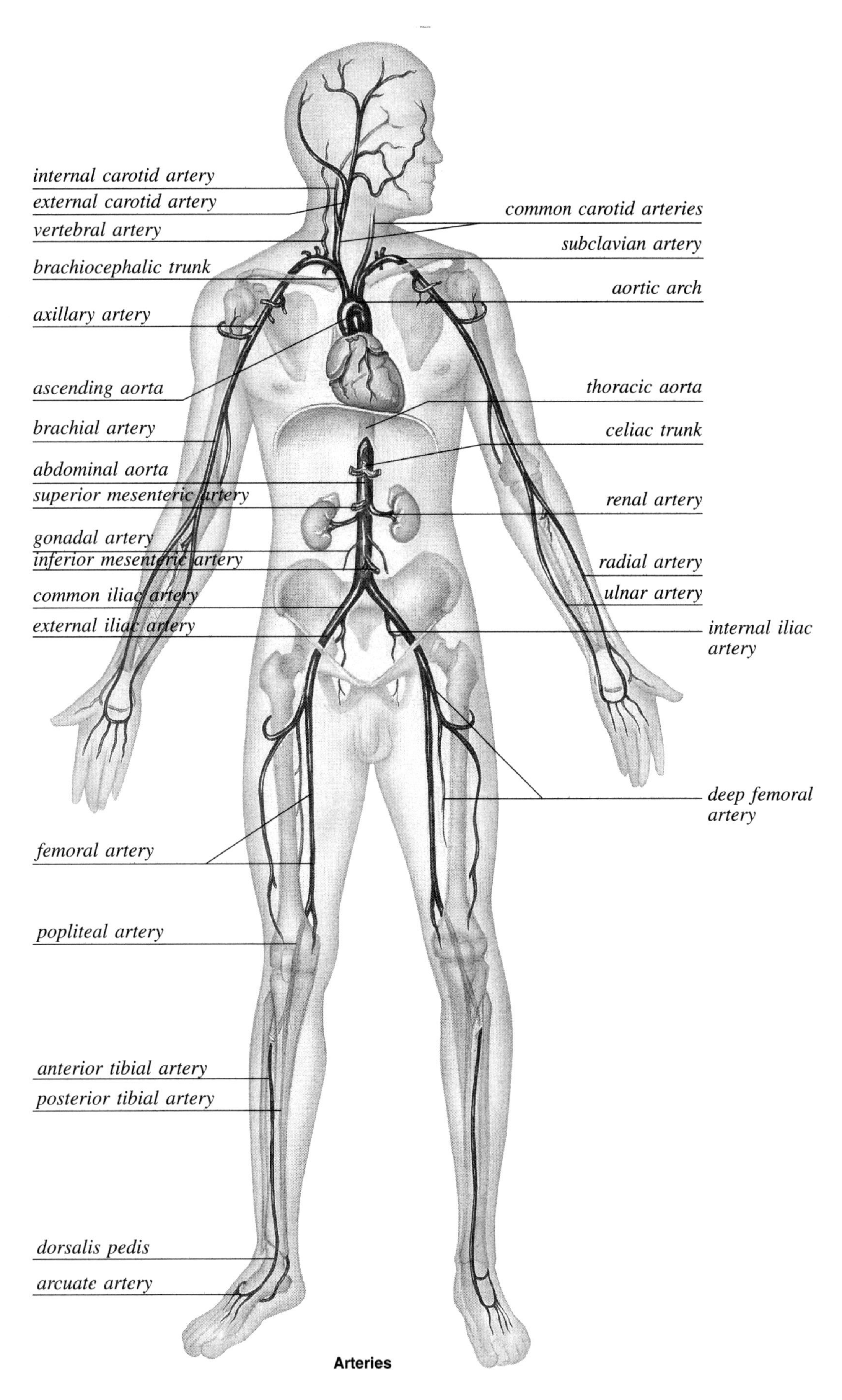

Arteries

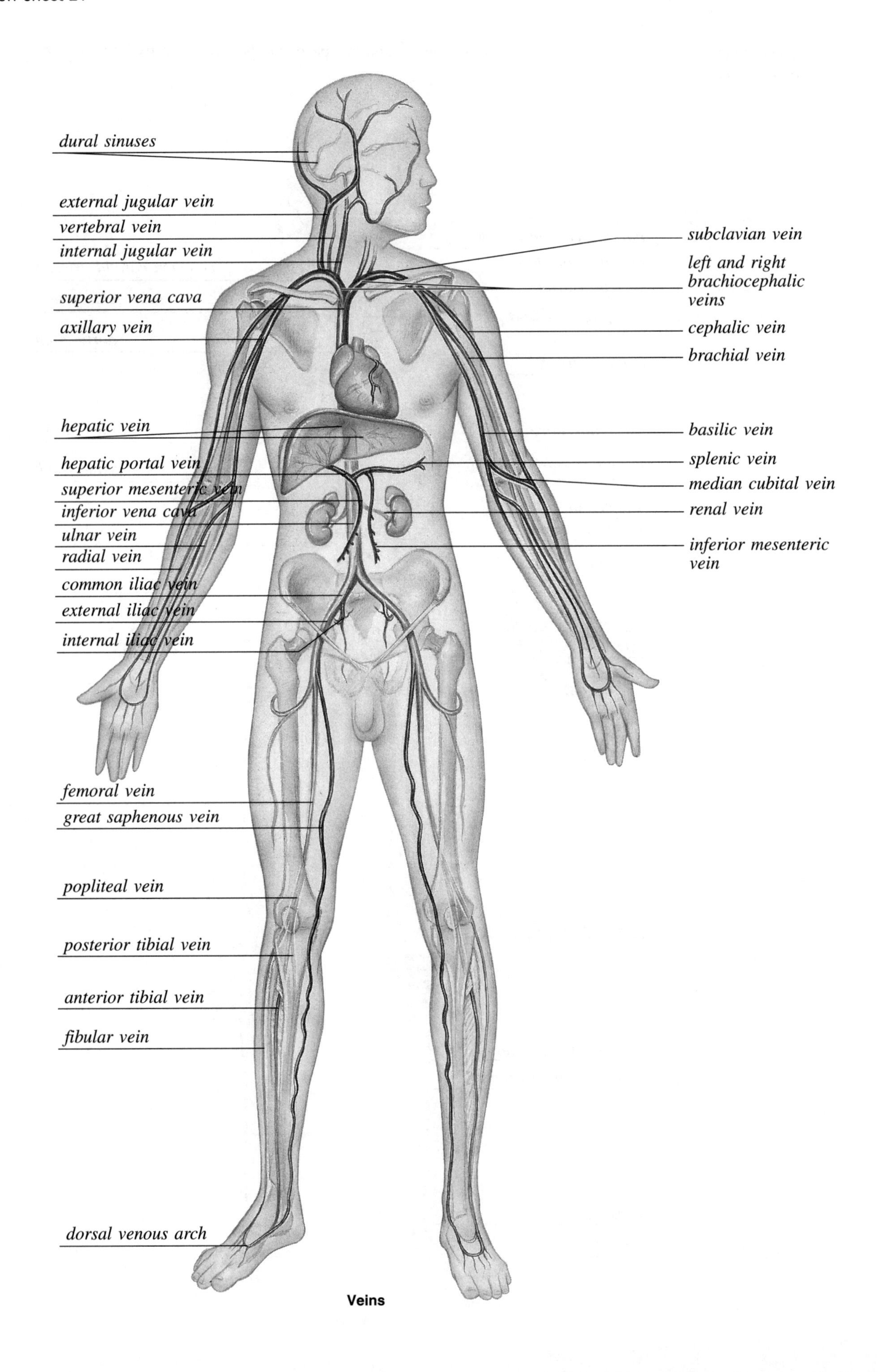
dural sinuses
external jugular vein
vertebral vein
internal jugular vein
superior vena cava
axillary vein
hepatic vein
hepatic portal vein
superior mesenteric vein
inferior vena cava
ulnar vein
radial vein
common iliac vein
external iliac vein
internal iliac vein
femoral vein
great saphenous vein
popliteal vein
posterior tibial vein
anterior tibial vein
fibular vein
dorsal venous arch
subclavian vein
left and right brachiocephalic veins
cephalic vein
brachial vein
basilic vein
splenic vein
median cubital vein
renal vein
inferior mesenteric vein

Veins

7. Trace the blood flow for the following situations:

a. From the capillary beds of the left thumb to the capillary beds of the right thumb *Capillaries → radial vein → brachial vein → axillary vein → left subclavian vein → brachiocephalic vein → superior vena cava → right atrium → right ventricle → pulmonary trunk → pulmonary arteries → lobar arteries → pulmonary capillaries → pulmonary venules → pulmonary veins → left atrium → left ventricle → aorta → brachiocephalic artery → right subclavian artery → axillary artery → brachial artery → radial artery*

b. From the pulmonary vein to the pulmonary artery by way of the right side of the brain *Pulmonary vein → left atrium left ventricle → aorta → brachiocephalic trunk → common carotid arteries → internal carotid artery → anterior and middle cerebral arteries → Circle of Willis → venous sinuses → internal jugular vein → right brachiocephalic vein → superior vena cava → right atrium → right ventricle → pulmonary artery*

Special Circulations

Pulmonary Circulation

8. Trace the pathway of a carbon dioxide gas molecule in the blood from the inferior vena cava until it leaves the bloodstream. Name all structures (vessels, heart chambers, and others) passed through en route.

Inferior vena cava → right atrium → right ventricle → pulmonary trunk → pulmonary arteries → lobar arteries → pulmonary capillaries → alveoli

9. Trace the pathway of an oxygen gas molecule from an alveolus of the lung to the right atrium of the heart. Name all structures through which it passes. *Alveolus pulmonary capillaries → pulmonary veins → left atrium → left ventricle → aorta → systemic arteries → arterioles → systemic capillaries → venules → veins → vena cava → right atrium*

10. Most arteries of the adult body carry oxygen-rich blood, and the veins carry oxygen-depleted, carbon dioxide–rich blood. What is different about the pulmonary arteries and veins? *Pulmonary veins carry oxygen-rich blood to the heart and pulmonary arteries carry carbon dioxide-rich blood to the lung.*

Hepatic Portal Circulation

11. What is the source of blood in the hepatic portal system? *The hepatic portal vein drains the digestive tract organs and carries this blood through the liver before it enters systemic circulation.*

12. Why is this blood carried to the liver before it enters the systemic circulation? *As blood goes through the liver, some of the nutrients are stored or processed in various ways for release to the general circulation.*

13. The hepatic portal vein is formed by the union of the *splenic vein*, which drains the *spleen*, *pancreas*, *stomach*, and the *large intestine* (via the interior mesenteric vein), and the *superior mesenteric vein*, which drains the *small intestine* and *the proximal colon*. The *left gastric vein* vein, which drains the lesser curvature of the stomach, empties directly into the hepatic portal vein.

14. Trace the flow of a drop of blood from the small intestine to the right atrium of the heart, noting all structures encountered or passed through on the way. *Small intestine → superior mesenteric vein → hepatic portal vein → inferior vena cava → right atrium*

Arterial Supply of the Brain and the Circle of Willis

15. Branches of the internal carotid and vertebral arteries cooperate to form a ring of blood vessels encircling the pituitary gland, at the base of the brain. What name is given to this communication network? *Circle of Willis*

    What is its function? *It is a protective device that provides an alternate set of pathways for blood to reach the brain tissue in case of impaired blood flow anywhere in the system.*

16. What portion of the brain is served by the anterior and middle cerebral arteries? *They supply the bulk of the cerebrum*

    Both the anterior and middle cerebral arteries arise from the *internal carotid* arteries.

17. Trace the usual pathway of a drop of blood from the aorta to the left occipital lobe of the brain, noting all structures through which it flows. Aorta → *subclavian arteries* → *vertebral arteries* → *basilar artery* → *posterior cerebral arteries* → left occipital lobe.

Fetal Circulation

18. The failure of two of the fetal bypass structures to become obliterated after birth can cause congenital heart disease, in which the youngster would have improperly oxygenated blood. Which two structures are these?

Ductus arteriosus and *Foramen ovale*

19. For each of the following structures, indicate its function in the fetus. Circle the blood vessel that carries the most oxygen-rich blood.

Structure	Function in fetus
Umbilical artery	*Carry CO_2 and waste-laden blood from the fetus to the placenta*
(Umbilical vein)	*Carries oxygen- and nutrient-rich blood from the placenta to the fetus*
Ductus venosus	*Allows blood to partially bypass the liver and carries blood to the right atrium of the heart*
Ductus arteriosus	*Bypasses the nonfunctional lungs and connects the pulmonary trunk and the aorta*
Foramen ovale	*Bypasses the nonfunctional lungs allowing blood from right atrium to move into the left atrium*

20. What organ serves as a respiratory/digestive/excretory organ for the fetus? *Placenta*

NAME ______________________________ LAB TIME/DATE ______________________

REVIEW SHEET
exercise
22

Human Cardiovascular Physiology—Blood Pressure and Pulse Determinations

Cardiac Cycle

1. Correctly identify valve closings and openings, chamber pressures, and volume lines, and the ECG and heart sound scan lines on the diagram below by using the terms from the list to the right of the diagram.

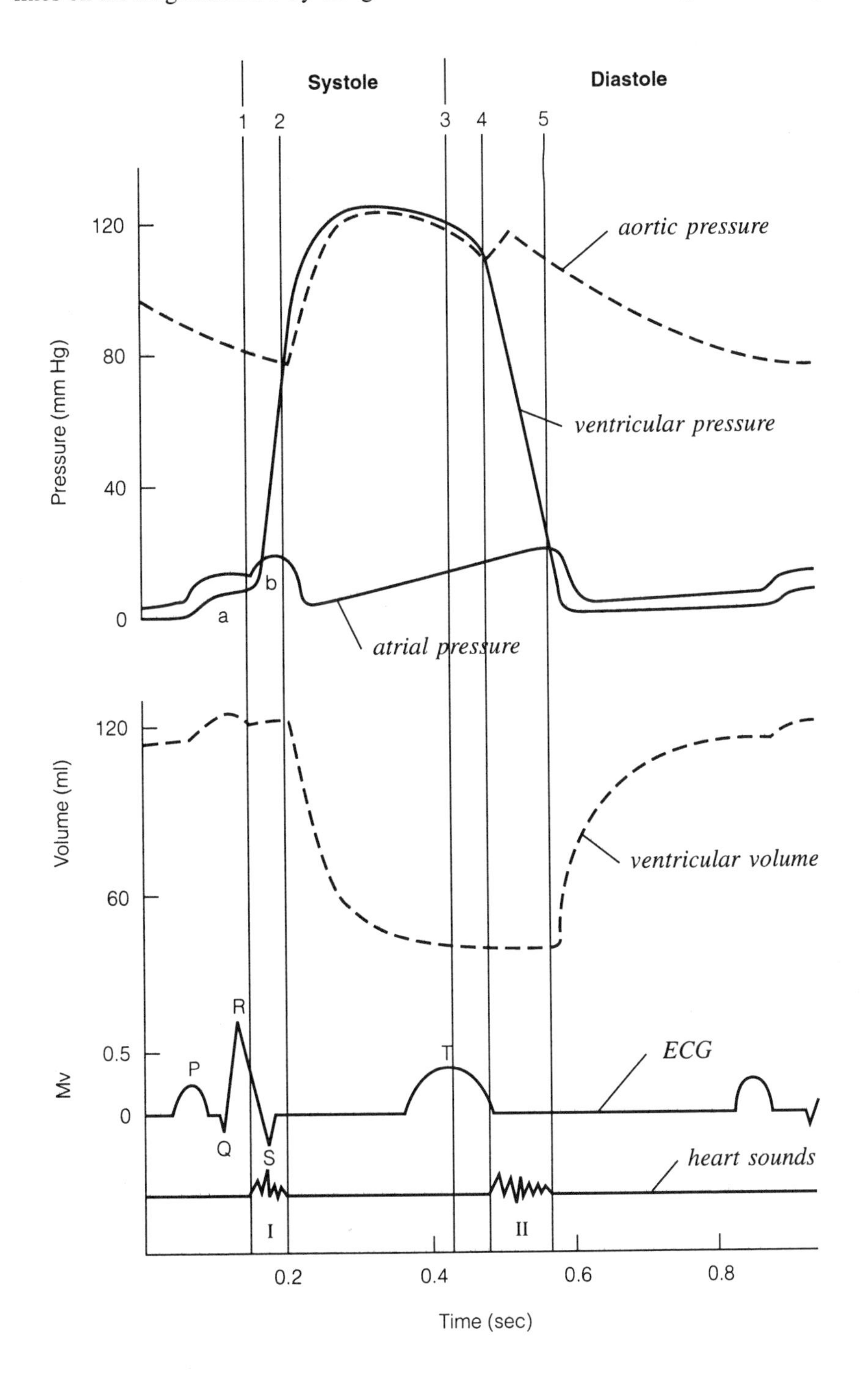

Key:

aortic pressure

atrial pressure

ECG

heart sounds

ventricular pressure

ventricular volume

2. Define the following terms:

Systole: *Contraction of the ventricles*

Diastole: *Relaxation of the ventricles*

Cardiac cycle: *Includes events of one complete heartbeat, during which both atria and ventricles contract and then relax*

3. Answer the following questions concerning events of the cardiac cycle:

When are the AV valves closed? *During ventricular systole*

Open? *During ventricular diastole*

What event within the heart causes the AV valves to open? *When ventricular pressure is less than atrial pressure, the AV valves are forced open*

What causes them to close? *Pressure in the ventricles rises rapidly*

When are the semilunar valves closed? *During ventricular diastole*

Open? *When ventricular systole occurs*

What event causes the semilunar valves to open? *The semilunar valves are forced open when ventricular pressure exceeds that of the large arteries leaving the heart.*

To close? *At the end of ventricular systole, the ventricles relax and the semilunar valves snap shut preventing backflow.*

At what point in the cardiac cycle is the pressure in the heart highest? *During ventricular systole*

Lowest? *During ventricular diastole*

4. If an individual's heart rate is 80 beats/min, what is the length of the cardiac cycle? *0.75 seconds*

Heart Sounds

5. Complete the following statements:

The monosyllables describing the heart sounds are __1__. The first heart sound is a result of closure of the __2__ valves, whereas the second is a result of closure of the __3__ valves. The heart chambers that have just been filled when you hear the first heart sound are the __4__, and the chambers that have just emptied are the __5__. Immediately after the second heart sound, the __6__ are filling with blood, and the __7__ are empty.

1. *"lub" "dup"*
2. *AV*
3. *semilunar*
4. *ventricles*
5. *atria*
6. *atria*
7. *ventricles*

6. As you listened to the heart sounds during the laboratory session, what differences in pitch, length, and amplitude (loudness) of the two sounds did you observe? *The first sound will be a longer, louder, more booming sound than the second, which is short and sharp.*

7. No one expects you to be a full-fledged physician on such short notice, but on the basis of what you have learned about heart sounds, how might abnormal sounds be used to diagnose heart problems? (Use your textbook as necessary.)

Abnormal sounds may be used to diagnose valve deformities that can seriously hamper cardiac function and ultimately weaken the heart. They may also be used to detect abnormalities in the conduction system of the heart and inadequate blood supply to the heart muscle.

The Pulse

8. Define *pulse:* *The alternating surges of pressure (expansion then recoil) in an artery that occur with each beat of the left ventricle*

9. Identify the artery palpated at each of the following pressure points:

at the wrist *radial artery*

on the dorsum of the foot *dorsalis pedis*

in front of the ear *temporal artery*

at the side of the neck *common carotid*

in the groin *femoral artery*

above the medial malleolus *posterior tibial*

10. How would you tell by simple observation whether bleeding is arterial *or* venous?

Arterial—bright red; in spurts   Venous—darker red; even flow

Blood Pressure Determinations

11. Define *blood pressure:* *The pressure the blood exerts against the inner blood vessel walls*

12. Identify the phase of the cardiac cycle to which each of the following apply:

systolic pressure *pressure in the arteries at the peak of ventricular ejection*

diastolic pressure *the pressure during ventricular relaxation*

13. What is the name of the instrument used to compress the artery and record pressures in the auscultatory method of determining blood pressure? *Sphygomomanometer*

14. What are sounds of Korotkoff? *Resumption of blood flow into the forearm, which sounds like soft tapping sounds*

What causes the systolic sound? *Resumption of blood flow into the forearm*

The disappearance of sound? *Blood flows freely because the artery is no longer compressed.*

15. Interpret 145/85. *Systolic pressure of 145 mm Hg; diastolic pressure of 85 mm Hg*

16. How do venous pressures compare to arterial pressures? *Venous pressure is lower than arterial.*

Why? *Because veins are farther away from the ventricular ejection*

Observing the Effect of Various Factors on Blood Pressure and Heart Rate

17. What effect do the following have on blood pressure? (Indicate increase by I and decrease by D.)

D 1. increased diameter of the arterioles

I 2. increased blood viscosity

I 3. increased cardiac output

D 4. hemorrhage

I 5. arteriosclerosis

I 6. increased pulse rate

18. In which position (sitting, reclining, or standing) is the blood pressure normally the highest?

Reclining The lowest? *Standing*

What immediate changes in blood pressure did you observe when the subject stood up after being in the sitting or reclining position? *Blood pressure dropped*

What changes in the blood vessels might account for the change? *Effect of gravity causes blood to pool in the vessels the legs and feet.*

After the subject stood for 3 minutes, what changes in blood pressure were observed? *Blood pressure came back to homeostasis.*

How do you account for this change? *Pressoreceptors send signal resulting in reflexive vasoconstriction.*

19. What was the effect of exercise on blood pressure? *Increased*

On pulse? *Increased* Do you think these effects reflect changes in cardiac output *or* in peripheral resistance? *Cardiac output*

Skin Color as an Indicator of Local Circulatory Dynamics

20. Describe normal skin color and the appearance of the veins in the subject's forearm before any testing was conducted.

Flesh tone, warm, dry

21. What changes occurred when the subject emptied the forearm of blood (by raising the arm and making a fist) and the flow was blocked with the cuff? *Pale, cold, clammy*

What changes occurred during venous congestion? *"Blue"ing, cold, swollen*

22. Explain the mechanism by which mechanical stimulation of the skin produced a flare. *Results from a local inflammatory response promoted by chemical mediators that stimulate increased blood flow into the area and cause the capillaries to leak fluid into the local tissues*

NAME ______________________ LAB TIME/DATE ______________________

REVIEW SHEET

exercise

23

Anatomy of the Respiratory System

Upper and Lower Respiratory System Structures

1. Complete the labeling of the diagram of the upper respiratory structures (sagittal section).

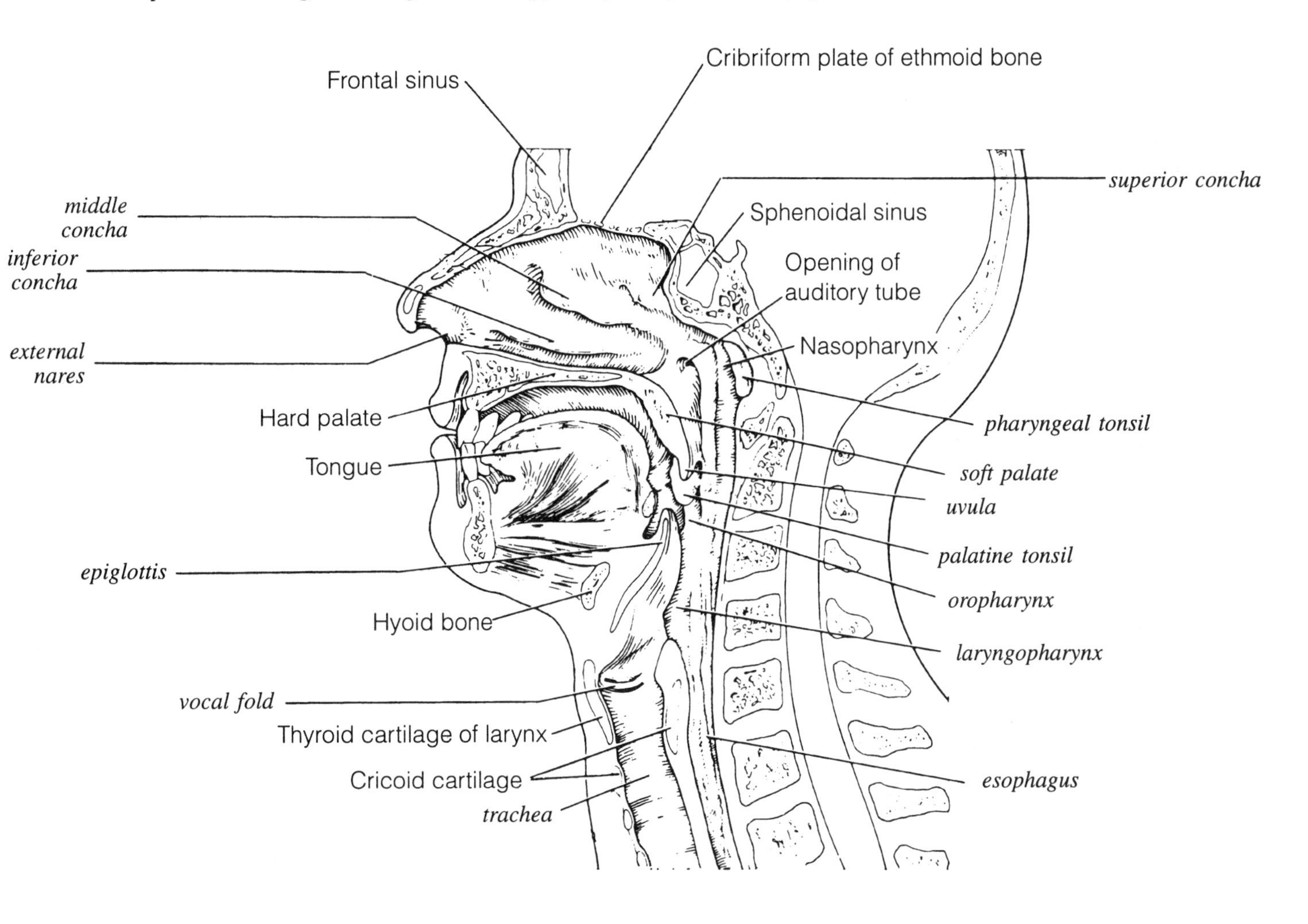

2. What is the significance of the fact that the human trachea is reinforced with cartilage rings?

Reinforce the trachea walls to keep its passageway open regardless of the pressure changes that occur during breathing.

Of the fact that the rings are incomplete posteriorly? *The open parts of these cartilage rings allow the esophagus to expand anteriorly when large pieces of food are swallowed.*

3. Name the specific cartilages in the larynx that are described below:

1. forms the Adam's apple *thyroid cartilage*
3. broader anteriorly *cricoid cartilage*
2. a "lid" for the larynx *epiglottis*

4. Trace a molecule of oxygen from the external nares to the pulmonary capillaries of the lungs:

External nares ⟶ *Nasal cavity → nasal conchae → nasopharynx → oropharynx → laryngopharynx → trachea → primary bronchus → secondary bronchus → tertiary bronchus → terminal bronchiole → respiratory bronchioles → alveolar duct → alveolus* ⟶ pulmonary capillaries

5. What is the function of the pleural membranes? *Produce lubricating serous fluid that causes them to adhere closely to one another, holding the lungs to the thoracic wall and allowing them to move easily against one another*

6. Name two functions of the nasal cavity mucosa: *Warmed, moistened, filtered*

7. The following questions refer to the primary bronchi:

Which is longer? *Left* Larger in diameter? *Right* More horizontal? *Left*

The more common site for lodging of a foreign object that has entered the respiratory passageways? *Right*

8. Appropriately label all structures provided with leader lines on the diagrams below.

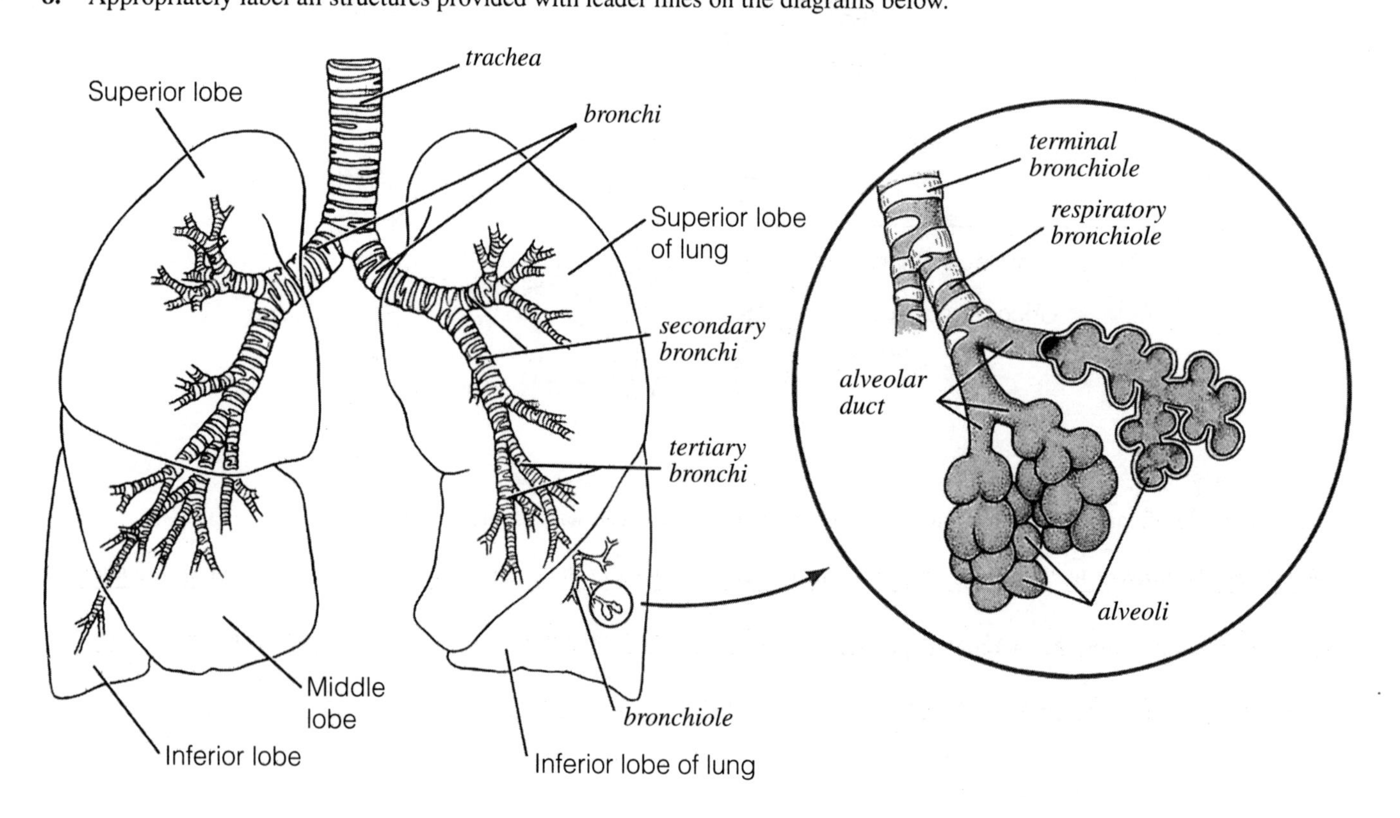

9. Match the terms in column B to those in column A.

	Column A	Column B
phrenic nerve	1. nerve that activates the diaphragm during inspiration	alveolus
palate	2. "floor" of the nasal cavity	bronchiole
esophagus	3. food and fluid passageway inferior to the laryngopharynx	concha
epiglottis	4. flaps over the glottis during swallowing of food	epiglottis
larynx	5. contains the vocal cords	esophagus
trachea	6. the part of the conducting pathway between the larynx and the primary bronchi	glottis
parietal pleura	7. pleural layer lining the walls of the thorax	larynx
alveolus	8. site from which oxygen enters the pulmonary blood	palate
glottis	9. opening between the vocal folds	parietal pleura
concha	10. increases air turbulence in the nasal cavity	phrenic nerve
		primary bronchi
		trachea
		visceral pleura

10. Define *external respiration:* *Gas exchanges to and from the pulmonary circuit blood that occur in the lungs*

internal respiration: *Exchange of gases to and from the blood capillaries of the systemic circulation*

Demonstrating Lung Inflation in a Sheep Pluck

11. Does the lung inflate part by part or as a whole, like a balloon? *The lung inflates part by part*

What happened when the pressure was released? *Lung deflates*

What type of tissue ensures this phenomenon? *Elastic connective tissue*

Examining Prepared Slides of Lung and Tracheal Tissue

12. The tracheal epithelium is ciliated and has goblet cells. What is the function of each of these modifications?

Cilia: *To push mucus and dust off of lungs*

Goblet cells: *Produce mucus*

13. The tracheal epithelium is said to be "pseudostratified." Why? *All of the cells of pseudostratified epithelium rest on a basement membrane; however, some of its cells are shorter than others, and their nuclei appear at different heights. As a result, the epithelium gives a false impression that it is stratified.*

14. On the diagram below identify aveolar epithelium, capillary endothelium, alveoli, and red blood cells. Bracket the respiratory membrane.

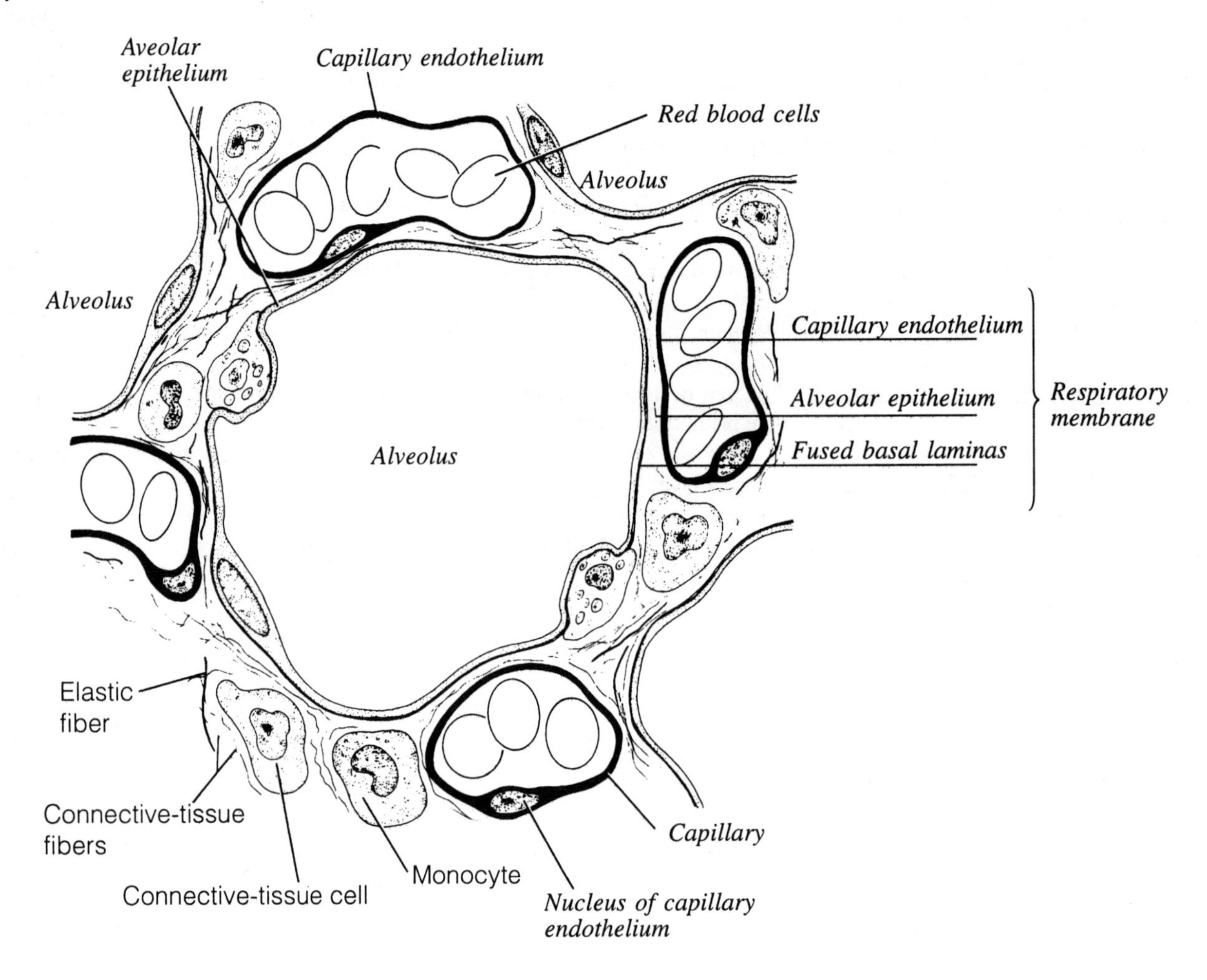

15. Why does oxygen move from the alveoli into the pulmonary capillary blood? *Due to simple diffusion, oxygen diffuses from the alveoli into the blood because the concentration of O_2 is higher in the alveoli.*

NAME ______________________ LAB TIME/DATE ______________________

REVIEW SHEET

exercise

24

Respiratory System Physiology

Mechanics of Respiration

1. Base your answers to the following on your observations of the operation of the model lung.

   Under what *internal* conditions does air tend to flow into the lungs? *The intrapulmonary volume increases, lowering the air pressure inside lungs.*

   Under what *internal* conditions does air tend to flow out of the lungs? Explain. *The intrapulmonary volume decreases, increasing the pressure inside the lungs.*

2. Activation of the diaphragm and the external intercostal muscles begins the inspiratory process. What results from the contraction of these muscles, and how is this accomplished? *The diaphragm contracts to a flattened position, increasing the superoinferior volume. The external intercostals lift the rib cage, increasing the anteroposterior and lateral dimensions.*

3. What was the approximate increase in diameter of chest circumference during a quiet inspiration?

   ______________ inches During forced inspiration? *(increased)* ______________ inches

   What temporary advantage does the substantial increase in chest circumference during forced inspiration create?

   The advantage of the chest circumference increase will allow an increase in intrapulmonary volume, lowering the air pressure inside the lungs, and gases then expand to fill the available space.

4. The presence of a partial vacuum between the pleural membranes is crucial to normal breathing movements. What would happen if an opening were made into the chest cavity, as with a puncture wound?

   An opening causes intrapleural pressure to equal atmospheric pressure and lungs to immediately recoil and collapse.

Respiratory Volumes and Capacities: Spirometry

5. Write the respiratory volume term and the normal value that is described by the following statements:

   Volume of air present in the lungs after a forceful expiration *Residual volume/1200 ml*

   Volume of air that can be expired forcibly after a normal expiration *Expiratory reserve volume/1000–1200 ml*

   Volume of air that is breathed in and out during a normal respiration *Tidal volume/500 ml*

Volume of air that can be inspired forcibly after a normal inspiration *Inspiratory reserve volume/2100–3100 ml*

Volume of air corresponding to TV + IRV + ERV *Vital capacity/3600–4800 ml*

6. Record experimental respiratory volumes as determined in the laboratory.

Average TV *500* ml Average VC *3600–4800* ml

Average ERV *1000–1200* ml Average IRV *2100–3100* ml

Factors Influencing Rate and Depth of Respiration

7. Where are the neural control centers of respiratory rhythm? *Medulla* and *Pons*

8. Based on pneumograph reading of respiratory variation, what was the rate of quiet breathing?

Initial testing *12–15* breaths/min

Record observations of how the initial pneumograph recording was modified during the various testing procedures described below. Indicate the respiratory rate, and include comments on the relative depth of the respiratory peaks observed.

Test performed	Observations
Talking	
Yawning	
Laughing	
Standing	
Concentrating	
Swallowing water	
Coughing	
Lying down	
Running in place	

9. Student data:

Breath-holding interval after a deep inhalation ______ sec length of recovery period ______ sec

Breath-holding interval after a forceful expiration ______ sec length of recovery period ______ sec

After breathing quietly and taking a deep breath (which you held), was your urge to inspire or expire? *Expire*

After exhaling and then holding your breath, did you want to inspire or expire? *Inspire*

Explain these results. (*Hint:* what reflex is involved here?) *Changes in O_2 and CO_2 concentrations in the blood act on the neural control centers of the respiratory rhythm. CO_2 changes act on medulla; O_2 levels are monitored by the chemoreceptors in the aortic and carotid bodies.*

10. Observations after hyperventilation: *Slower breathing rate; shallow*

11. Length of breath holding after hyperventilation: ________ sec

Why does hyperventilation produce apnea or a reduced respiratory rate? *Carbonic acid decreases greatly during hyperventilation, increasing (alkalosis) the pH. Apnea will occur until CO_2 builds up again in the blood.*

12. Observations for rebreathing breathed air: *Breathing rate becomes rapid and deep.*

Why does rebreathing breathed air produce an increased respiratory rate? *CO_2 levels are increased in blood leading to increased breathing rate.*

13. What was the effect of running in place (exercise) on the duration of breath holding? *Shortens duration*

Explain: *Respiratory centers ignore voluntary control when CO_2 level is increased in blood or pH level is decreased*

14. Record your data for the test illustrating the effect of respiration on circulation:

Radial pulse before beginning test ________ /min Radial pulse after testing *(slower)* /min

Relative radial pulse force before beginning test ________ Relative radial pulse force after testing *(weaker)*

Condition of neck and facial veins after testing *extended/protruding*

Explain: *CO_2 build-up was expelled during forced expiration. Until $\uparrow CO_2$ level/pH level$\downarrow$ there is no stimulus to increase heart rate. Blood vessels dilate to allow more blood passage and heat release.*

15. Do the following factors generally increase (indicate with I) or decrease (indicate with D) the respiratory rate and depth?

1. increase in blood CO_2 *I*
2. decrease in blood O_2 *I*
3. increase in blood pH *D*
4. decrease in blood pH *I*

Did it appear that CO_2 or O_2 had a greater effect on modifying the respiratory rate? *CO_2*

16. Where are sensory receptors sensitive to changes in O_2 levels in the blood located? *In the aorta and carotid artery*

17. Blood CO_2 levels and blood pH are related. When blood CO_2 levels increase, does the pH increase or decrease?

Decreases Explain why. *CO_2 retention (increased levels) leads to increased levels of carbonic acid, which decreases the blood pH.*

NAME ______________________ LAB TIME/DATE ______________________

REVIEW SHEET
exercise
25

Functional Anatomy of the Digestive System

General Histological Plan of the Alimentary Canal

1. The basic structural plan of the digestive tube has been presented. Fill in the table below to complete the information listed.

Wall layer	Subdivisions of the layer	Major functions
mucosa	• *Surface epithelium* • *Lamina propria* • *Muscular layer*	• *Secretion of enzymes, mucus, hormones, etc.* • *Absorption of digested foodstuffs* • *Protection against bacterial invasion*
submucosa	• *Connective tissue* • *Lymph nodules* • *Nerve fibers*	• *Nutrition* • *Protection*
muscularis externa	• *Circular muscle layer* • *Longitudinal muscle layer*	• *Regulates GI motility*
serosa	*Visceral peritoneum*	• *Reduces friction as the GI tract organs work*

Organs of the Alimentary Canal

2. The tubelike digestive system canal that extends from the mouth to the anus is the *alimentary* canal.

3. How is the muscularis externa of the stomach modified? *It contains a third obliquely oriented layer of smooth muscle in its muscularis externa.*

How does this modification relate to the stomach's function? *It allows the stomach to churn, mix, and pummel the food, physically breaking it down to smaller fragments.*

4. Using the key letters, match the items in column B with the descriptive statements in column A.

Column A

k ______ 1. structure that suspends the small intestine from the posterior body wall

l ______, y ______, q ______ 2. three modifications of the small intestine that increase the surface area for absorption

o ______ 3. large collections of lymphoid tissue found in the submucosa of the small intestine

q ______ 4. deep folds of the mucosa and submucosa that extend completely or partially around the circumference of the small intestine

m ______, v ______ 5. regions that break down foodstuffs mechanically

w ______ 6. mobile organ that initiates swallowing

p ______ 7. conduit that serves the respiratory and digestive systems

c ______ 8. the "gullet"; lies posterior to the trachea

l ______ 9. surface projections of a mucosal epithelial cell

h ______ 10. valve at the junction of the small and large intestines

t ______ 11. primary region of enzymatic digestion

d ______ 12. membrane securing the tongue to the floor of the mouth

x ______ 13. area between the teeth and lips/cheeks

b ______ 14. wormlike sac that outpockets from the cecum

m ______ 15. carbohydrate (starch) digestion begins here

e ______ 16. two-layered serous membrane attached to the greater curvature of the stomach

i ______ 17. organ distal to the small intestine

r ______ 18. valve preventing movement of chyme from the duodenum into the stomach

u ______ 19. posterosuperior boundary of the oral cavity

t ______ 20. location of the hepatopancreatic sphincter through which pancreatic secretions and bile pass

z ______ 21. outermost layer of a digestive organ in the abdominal cavity

i ______ 22. principal site for the synthesis of vitamins (B, K) by bacteria

a ______ 23. distal end of the alimentary canal

f ______ 24. bone-supported part of roof of the mouth

Column B

a. anus
b. appendix
c. esophagus
d. frenulum
e. greater omentum
f. hard palate
g. haustra
h. ileocecal valve
i. large intestine
j. lesser omentum
k. mesentery
l. microvilli
m. oral cavity
n. parietal peritoneum
o. Peyer's patches
p. pharynx
q. plicae circulares
r. pyloric valve
s. rugae
t. small intestine
u. soft palate
v. stomach
w. tongue
x. vestibule
y. villi
z. visceral peritoneum

5. Correctly identify all structures depicted in the diagram below.

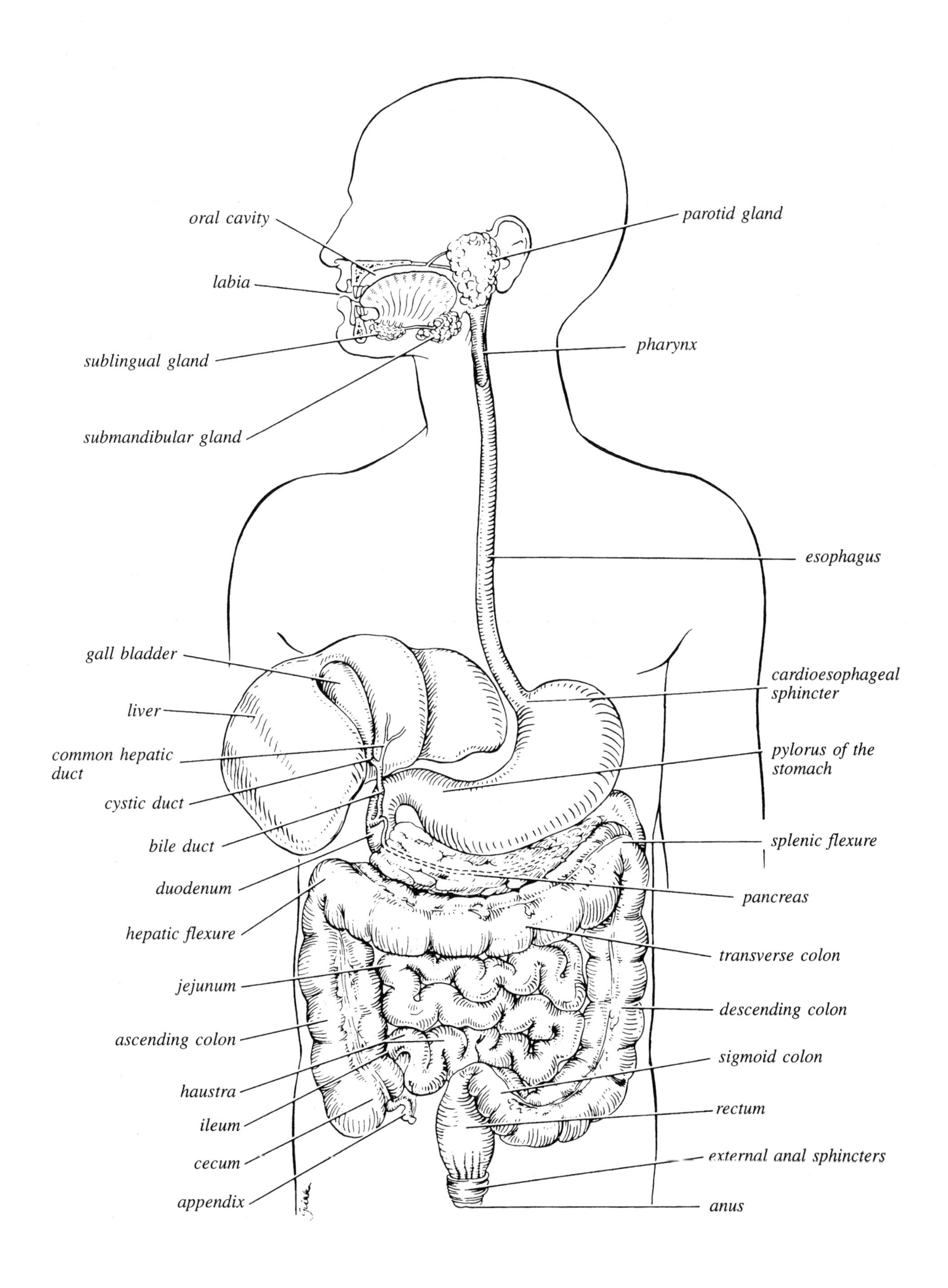

Accessory Digestive Organs

6. Use the key terms to identify each tooth area described below.

Answer	Description	Key:
crown	1. visible portion of the tooth	cementum
cementum	2. material covering the tooth root	crown
enamel	3. hardest substance in the body	dentin
periodontal ligament	4. attaches the tooth to bone and surrounding alveolar structures	enamel
root	5. portion of the tooth embedded in bone	gingiva
dentin	6. forms the major portion of tooth structure; similar to bone	periodontal ligament
pulp	7. produces the dentin	pulp
pulp	8. site of blood vessels, nerves, and lymphatics	root
crown	9. portion of the tooth covered with enamel	

7. In humans, the number of deciduous teeth is *20*; the number of permanent teeth is *32*.

8. The dental formula for permanent teeth is $\frac{2, 1, 2, 3}{2, 1, 2, 3}$

Explain what this means: *Upper teeth: 2 incisors, 1 canine, 2 premolars, 3 molars*

Lower teeth: 2 incisors, 1 canine, 2 premolars, 3 molars

9. What teeth are the "wisdom teeth"? *The number 3 molars*

10. Various types of glands form a part of the alimentary tube wall or release their secretions into it by means of ducts. Match the glands listed in column B with the function/locations described in column A.

Answer	Column A	Column B
duodenal glands	1. produce(s) mucus; found in the submucosa of the small intestine	duodenal glands
salivary glands	2. produce(s) a product containing amylase that begins starch breakdown in the mouth	gastric glands
pancreas	3. produce(s) a whole spectrum of enzymes and an alkaline fluid that is secreted into the duodenum	liver
liver	4. produce(s) bile that it secretes into the duodenum via the bile duct	pancreas
gastric glands	5. produce(s) HCl and pepsinogen	salivary glands

11. What is the role of the gallbladder? *Bile is stored there until needed for the digestive process.*

Chemical Digestion of Foodstuffs: Enzymatic Action

12. Match the following definitions with the proper choices from the key.

Key: catalyst, control, enzyme, substrate

catalyst 1. increases the rate of a chemical reaction without becoming part of the product

control 2. provides a standard of comparison for test results

enzyme 3. biologic catalyst: protein in nature

substrate 4. substance on which a catalyst works

13. The enzymes of the digestive system are classified as hydrolases. What does this mean?

They break down substrates by adding water to the molecular bonds, thus breaking the bonds between the monomers.

14. Fill in the following chart about the various digestive system enzymes described in this exercise.

Enzyme	Organ producing it	Site of action	Substrate(s)	Optimal pH
Salivary amylase	*Salivary glands*	*Oral cavity*	*Starch*	*Neutral*
Trypsin	*Pancreas*	*Small intestine*	*Protein*	*Alkaline*
Lipase (pancreatic)	*Pancreas*	*Small intestine*	*Fats*	*Alkaline*

15. Name the end products of digestion for the following types of foods:

proteins: *amino acids* carbohydrates: *glucose*

fats: *glycerol* and *fatty acids*

16. In the exercise concerning trypsin function, how could you tell protein hydrolysis occurred? *Appearance of a yellow color*

Why was tube 1T necessary? *Control tube of trypsin to compare against experimental sample*

Why was tube 2T necessary? *Control tube of BAPNA to compare against experimental sample*

Why was 37°C the optimal incubation temperature? *37°C is normal body temperature*

Why did very little, if any, digestion occur in test tube 3T? *Boiling the enzyme denatured it and rendered it useless.*

Why did very little, if any, digestion occur in test tube 5T? *The 0° temperature affected the enzyme, inhibiting its activity.*

Trypsin is a protein digesting enzyme similar to pepsin, the protein-digesting enzyme in the stomach. Would trypsin work well in the stomach? *No* Why? *The stomach has an acidic pH and trypsin works in an alkaline pH.*

17. In the procedure concerning the action of bile salts, how did the appearance of tubes 1 and 2 differ? *In tube 1 the oil was floating on the surface of the water. In tube 2 it appeared that the fat droplets were suspended in the water.*

Explain the difference. *Tube 2 had bile salts that emulsified the fat.*

18. Pancreatic and intestinal enzymes operate optimally at a pH that is slightly alkaline, yet the chyme entering the duodenum from the stomach is very acid. How is the proper pH for the functioning of the pancreatic-intestinal enzymes ensured?

A high concentration of bicarbonate ion (HCO_3^-) neutralizes the acidic chyme entering the duodenum from the stomach, enabling the pancreatic and intestinal enzymes to operate at their optimal pH.

19. Assume you have been chewing a piece of bread for 5 or 6 minutes. How would you expect its taste to change during this interval? *The bread will begin to taste sweet.*

Why? *Salivary amylase will break down the starch into sucrose.*

20. In the space below, draw the pathway of a ham sandwich (ham = protein and fat; bread = starch) from the mouth to the site of absorption of its breakdown products, noting where digestion occurs and what specific enzymes are involved.

See Table 25.1 in the laboratory manual, page 223

Physical Processes: Mechanisms of Food Propulsion and Mixing

21. Match the items in the key to the descriptive statements that follow.

uvula	1. blocks off nasal passages during swallowing
buccal	2. voluntary phase of swallowing
peristalsis	3. propulsive waves of smooth muscle contraction
cardioesophageal	4. sphincter that opens when food or fluids exert pressure on it
segmental	5. movement that mainly serves to mix foodstuffs
tongue	6. forces food into the pharynx
pharyngeal–esophageal	7. involuntary phase of swallowing

Key:
- buccal
- cardioesophageal
- peristalsis
- pharyngeal–esophageal
- segmental
- tongue
- uvula

NAME________________________ LAB TIME/DATE ________________

REVIEW SHEET
exercise
26

Functional Anatomy of the Urinary System

Gross Anatomy of the Human Urinary System

1. What is the function of the fat cushion that surrounds the kidneys in life? *In a living person, fat deposits (adipose capsules) hold the kidneys in place against the muscles of the posterior trunk wall.*

2. Complete the labeling of the diagram to correctly identify the urinary system organs. Then respond to the questions that follow.

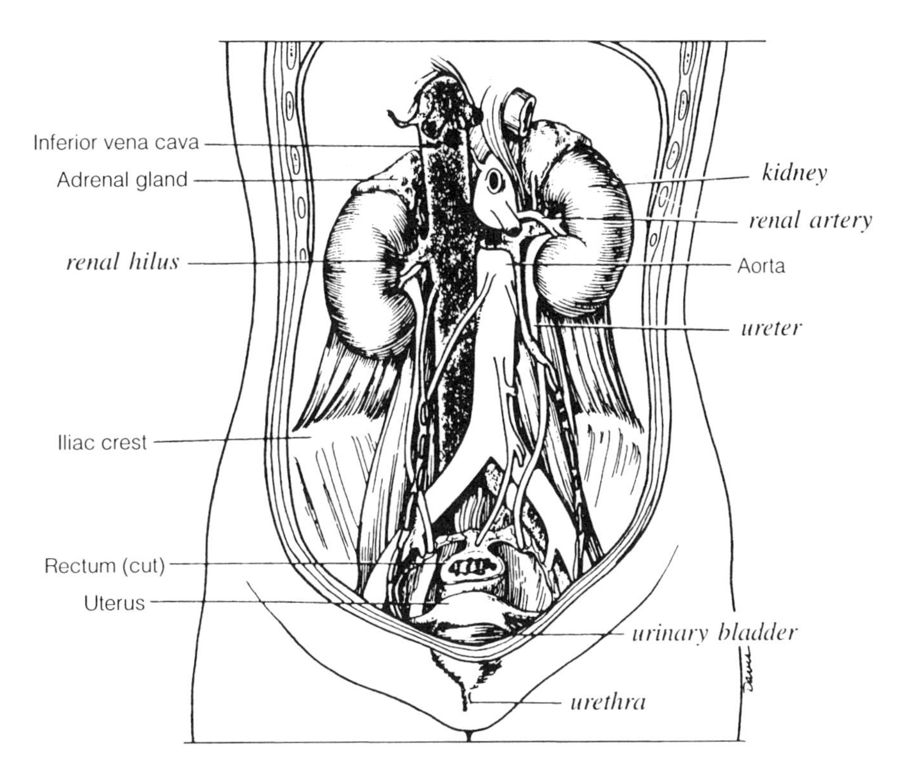

Which of the identified organs does the following?

kidney ________ 1. maintains water and electrolyte balance of the blood

urinary bladder ________ 2. serves as a storage area for urine

urethra ________ 3. transports urine to the body exterior

renal artery ________ 4. transports arterial blood to the kidney

kidney ________ 5. produces urine

ureter ________ 6. transports urine to the urinary bladder

urethra ________ 7. is shorter in females than in males

Gross Internal Anatomy of the Pig or Sheep Kidney

3. Match the appropriate structure in column B to its description in column A.

	Column A	Column B
renal capsule	1. smooth membrane clinging tightly to the kidney surface	cortex
medulla	2. portion of the kidney containing mostly collecting ducts	medulla
cortex	3. portion of the kidney containing the bulk of the nephron structures	calyx
cortex	4. superficial region of kidney tissue	renal capsule
renal pelvis	5. basinlike area of the kidney, continuous with the ureter	renal column
calyx	6. an extension of the pelvis that encircles the apex of a pyramid	renal pelvis
renal column	7. area of cortexlike tissue running between the medullary pyramids	

Functional Microscopic Anatomy of the Kidney

4. Match each of the lettered structures on the diagram of the nephron (and associated renal blood supply) on the left with the terms on the right:

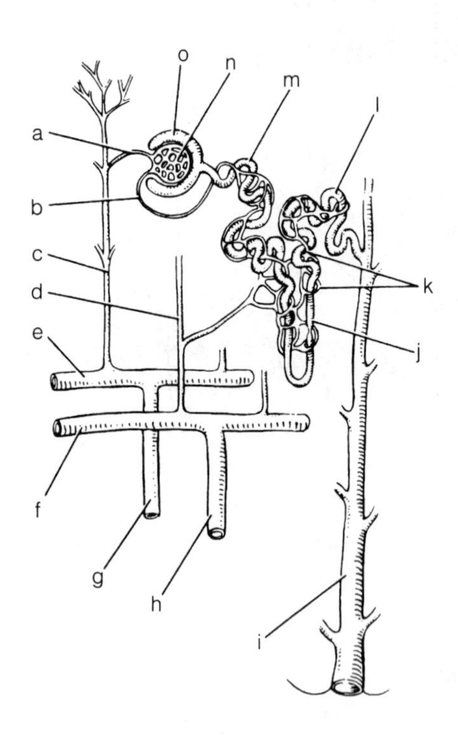

i 1. collecting duct

n 2. glomerulus

k 3. peritubular capillaries

l 4. distal convoluted tubule

m 5. proximal convoluted tubule

g 6. interlobar artery

c 7. interlobular artery

e 8. arcuate artery

d 9. interlobular vein

b 10. efferent arteriole

f 11. arcuate vein

j 12. loop of Henle

a 13. afferent arteriole

h 14. interlobar vein

o 15. glomerular capsule

5. Using the terms provided in question 1, identify the following:

glomerulus 1. site of filtrate formation

proximal convoluted tubule 2. primary site of tubular reabsorption

collecting duct 3. structure that conveys the processed filtrate (urine) to the renal pelvis

peritubular capillaries 4. blood supply that directly receives substances from the tubular cells

glomerulus 5. its inner (visceral) membrane forms part of the filtration membrane

6. Explain *why* the glomerulus is such a high-pressure capillary bed. *The glomerulus is fed and drained by arterioles and the feeder afferent arteriole is larger in diameter than the efferent arteriole draining the bed.*

How does its high-pressure condition aid its function of filtrate formation? *The high hydrostatic pressure forces out fluid and blood components smaller than proteins from the glomerulus into the glomerular capsule.*

7. What structural modification of certain tubule cells enhances their ability to reabsorb substances from the filtrate?

The inner wall of the glomerular capsule consists of specialized cells with long branching processes called podocytes.

8. Trace a drop of blood from the time it enters the kidney in the renal artery until it leaves the kidney through the renal vein.

Renal artery → *segmental artery → lobar artery → interlobar artery → arcuate artery → interlobular artery → afferent arterioles → glomerulus → peritubular capillary → interolobular vein → arcuate vein → interlobar vein* → renal vein

9. Trace the anatomical pathway of a molecule of creatinine (metabolic waste) from the glomerular capsule to the urethra. Note each microscopic and/or gross structure it passes through in its travels, and include the names of the subdivisions of the renal tubule.

Glomerular capsule → *proximal convoluted tubule → loope of Henle → distal convoluted tubule → collecting duct → renal cortex → medullary pyramid → calyces → renal pelvis → ureter → urinary bladder* → urethra

Urinalysis: Characteristics of Urine

10. What is the normal volume of urine excreted in a 24-hour period? *1.0 to 1.8 liters*

11. List three nitrogenous wastes that are routinely found in urine:

Urea, creatinine, uric acid

List three substances that are absent from the filtrate *and* urine of healthy individuals:

Blood, protein, bile

List two substances that are routinely found in filtrate but not in the urine product:

Glucose, amino acids

12. Explain why urinalysis is a routine part of any good physical examination. *It demonstrates kidney function, which maintains the electrolyte, acid-base, and fluid balances of the blood.*

13. What substance is responsible for the normal yellow color of urine? *Urochrome (breakdown of hemoglobin)*

14. Which has a greater specific gravity: 1 ml of urine or 1 ml of distilled water? *1 ml of urine*

    Explain. *Because urine contains dissolved solutes, it weighs more than water*

15. Explain the relationship between the color, specific gravity, and volume of urine. *The greater the solute concentration, the deeper the yellow color, the higher the specific gravity, and the smaller the output.*

Abnormal Urinary Constituents

16. How does a urinary tract infection influence urine pH? *A bacterial infection may result in urine with a high pH.*

    How does starvation influence urine pH? *Protein (in muscle breakdown) increases the acidity of urine. Elevated levels of ketones in urine = acid urine = decreased pH*

17. Several specific terms have been used to indicate the presence of abnormal urine constituents. Identify which urine abnormalities listed in Column A might be caused by each of the conditions described below by inserting a term from the list in Column B.

	Column A	Column B
hematuria	1. blood in the urine	albuminuria
hemoglobinuria	2. hemolytic anemia	glycosuria
glycosuria	3. eating a 5-lb box of candy at one sitting	hematuria
albuminuria	4. pregnancy	hemoglobinuria
ketonuria	5. starvation	ketonuria
pyouria	6. urinary tract infection	pyuria

18. What are renal calculi, and what conditions favor their formation? *Renal calculi are kidney stones that may form if the urine becomes excessively concentrated and solutes begin to precipitate or crystallize.*

19. All urine specimens become alkaline and cloudy on standing at room temperature. Explain. *Bacteria will grow and break down urea to form ammonia.*

NAME ______________________ LAB TIME/DATE ______________

REVIEW SHEET
exercise
27

Anatomy of the Reproductive System

Gross Anatomy of the Human Male Reproductive System

1. List the two principal functions of the testis: *Sperm production, testosterone production*

2. Identify all indicated structures or portions of structures on the diagrammatic view of the male reproductive system below.

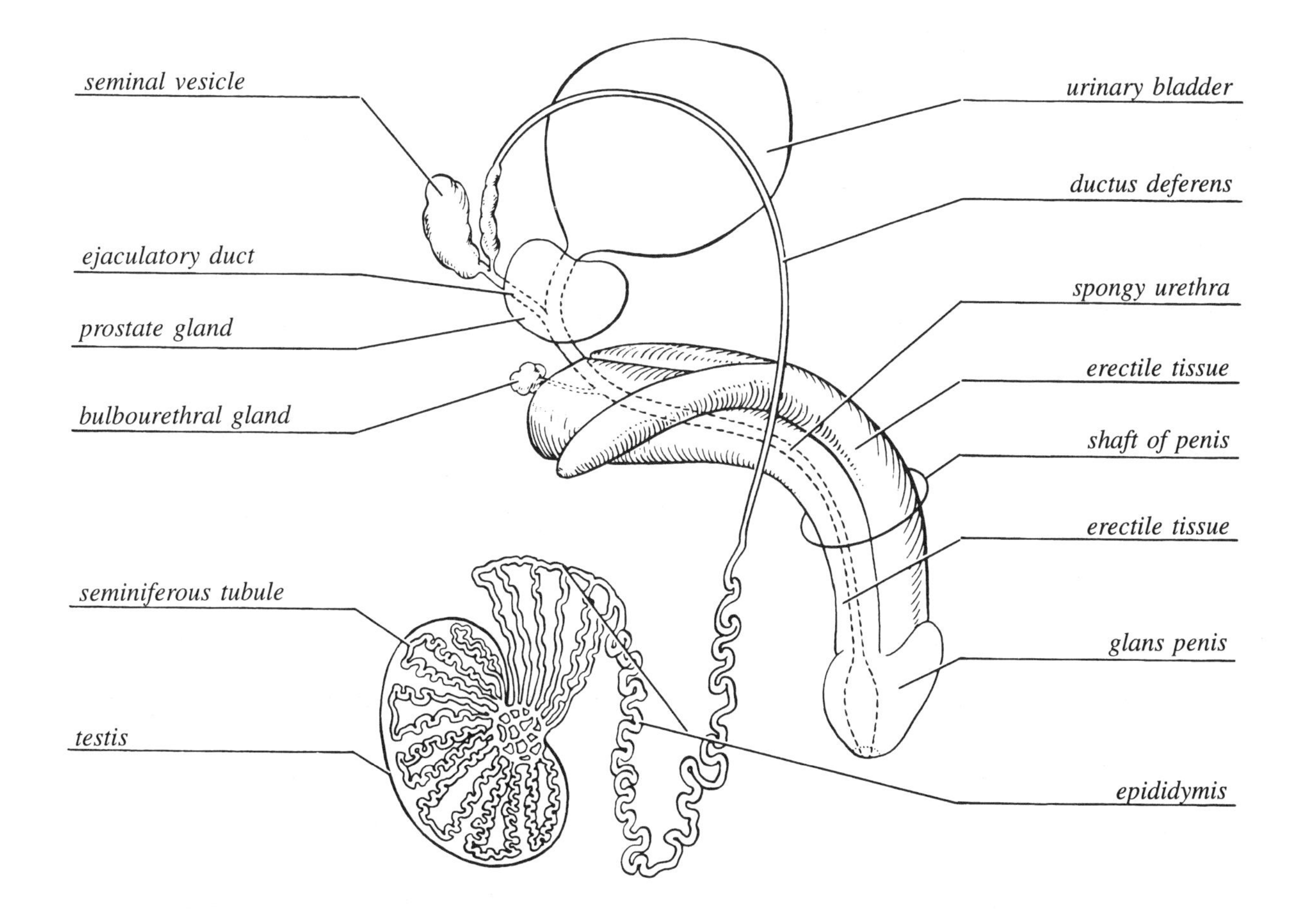

3. How might enlargement of the prostate gland interfere with urination or the reproductive ability of the male?

The prostate gland encircles the urethra just inferior to the bladder. If the gland is enlarged, it will constrict the urethra.

4. Match the terms in column B to the descriptive statements in column A.

	Column A	Column B
penis	1. copulatory organ/penetrating device	bulbourethral gland
testis	2. produces sperm	epididymis
vas deferens	3. duct conveying sperm to the ejaculatory duct; in the spermatic cord	glans penis
		membranous urethra
penile urethra	4. a urine and semen conduit	penile urethra
epididymis	5. sperm maturation site	penis
scrotum	6. location of the testis in adult males	prepuce
prepuce	7. hoods the glans penis	prostate gland
membranous urethra	8. portion of the urethra between the prostate gland and the penis	prostatic urethra
prostate gland	9. empties a secretion into the prostatic urethra	seminal vesicle
bulbourethral gland	10. empties a secretion into the membranous urethra	scrotum
		testis
		ductus deferens

5. Why are the testes located in the scrotum? *The temperature there is slightly lower than body temperature, a requirement for producing viable sperm.*

6. Describe the composition of semen and name all structures contributing to its formation. *Semen is composed of sperm and seminal fluid. The sperm are produced in the testis and the seminal fluid is produced by the prostate, seminal vesicles, and bulbourethral gland.*

7. Of what importance is the fact that seminal fluid is alkaline? *The relative alkalinity of semen helps neutralize the acid environment of the female vagina, protecting the sperm, and enhancing their motility.*

8. Using the following terms, trace the pathway of sperm from the testes to the urethra: rete testis, epididymis, seminiferous tubule, ductus deferens.

seminiferous tubule → *rete testis* → *epididymis* → *ductus deferens*

Gross Anatomy of the Human Female Reproductive System

9. On the diagram of a frontal section of a portion of the female reproductive system seen on the following page, identify all indicated structures.

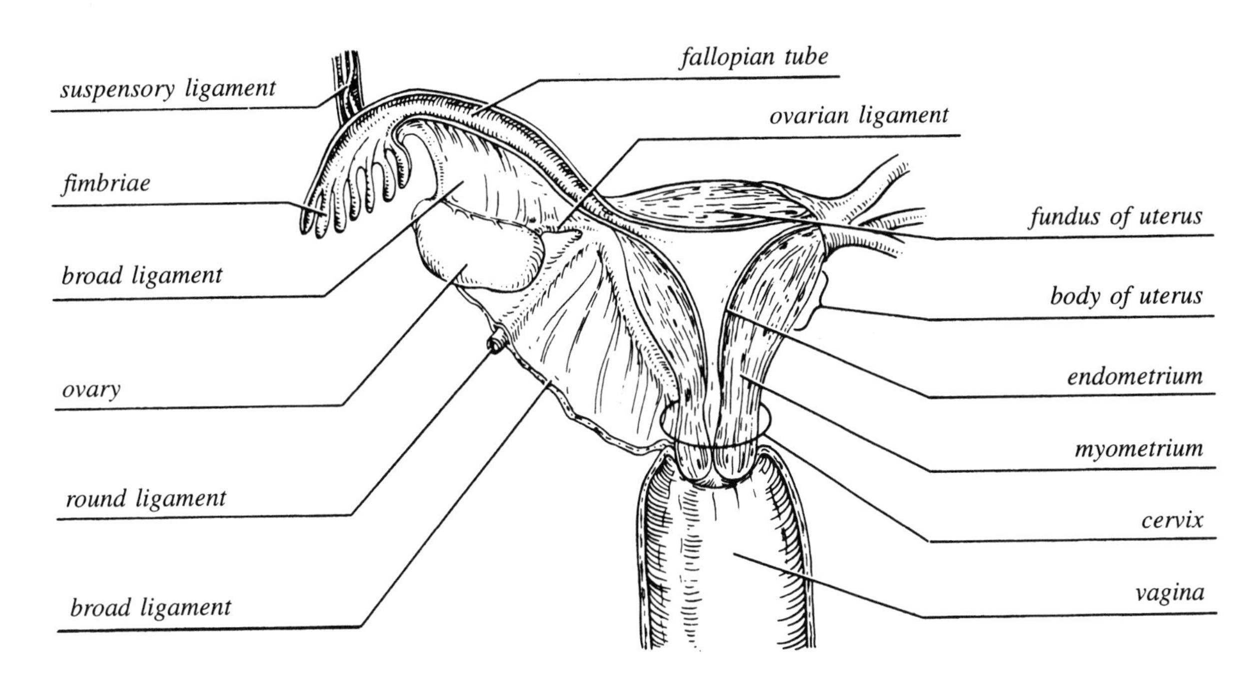

10. Identify the female reproductive system structures described below:

uterus 1. site of fetal development

vagina 2. copulatory canal

fallopian tube 3. "fertilized egg" typically formed here

clitoris 4. becomes erectile during sexual excitement

fallopian tubes 5. duct extending superolaterally from the uterus

ovary 6. produces eggs, estrogens, and progesterone

fimbriae 7. fingerlike ends of the uterine tube

11. Do any sperm enter the pelvic cavity of the female? Why or why not? *Possibly, because there is no actual contact between the female gonad and the uterine tube*

12. Name the structures composing the external genitals, or vulva, of the female. *The external genitals consist of the mons pubis, the labia majora and minora, the clitoris, the urethral and vaginal orifices, and greater vestibular glands.*

13. Put the following vestibular-perineal structures in their proper order from the anterior to the posterior aspect: vaginal orifice, anus, urethral opening, and clitoris.

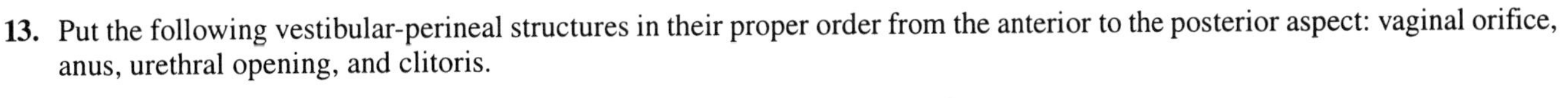
Anterior limit: *clitoris* → *urethral opening* → *vaginal orifice* → *anus*

14. Name the male structure that is homologous to the female structures named below.

labia majora *scrotum* clitoris *penis*

15. Assume a couple has just consummated the sex act and the male's sperm have been deposited in the woman's vagina. Trace the pathway of the sperm through the female reproductive tract.

vagina → cervix → uterus → fallopian tube

16. Define *ovulation:* *The ejection of a mature egg from the ovary*

Microscopic Anatomy of Selected Male and Female Reproductive Organs

17. The testis is divided into a number of lobes by connective tissue. Each of these lobes contains one to four *highly coiled seminiferous tubules*, which converge on a tubular region of the testis called the *rete testis at the mediastinum of the testis.*

18. What is the function of the spongy erectile bodies seen in the male penis? *These spongy erectile bodies fill with blood during sexual excitement, causing the penis to enlarge and become rigid.*

19. On the diagram showing the sagittal section of the human testis, correctly identify all structures provided with leader lines.

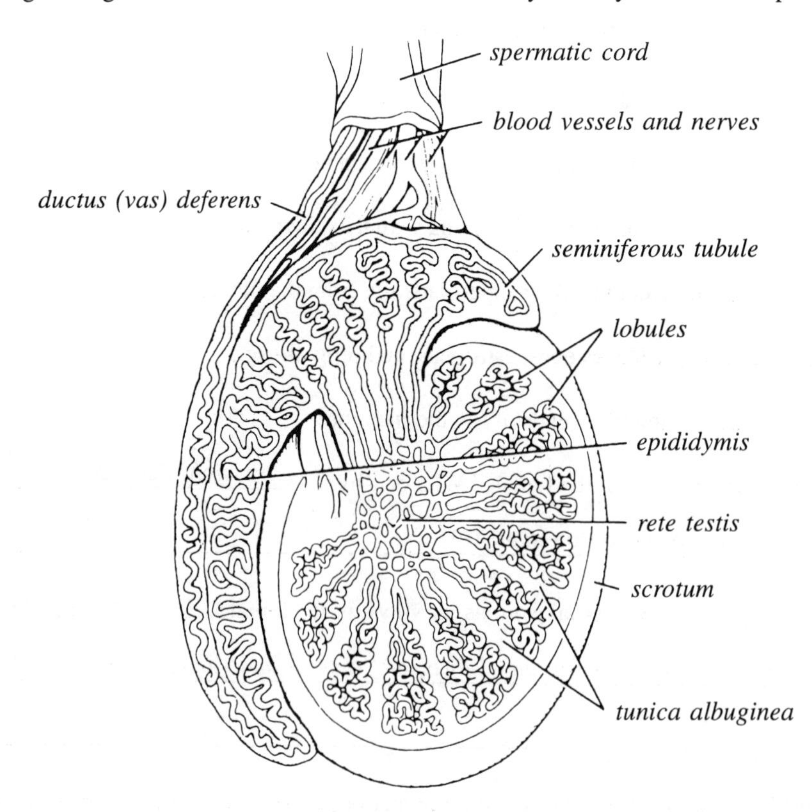

20. The female gametes develop in structures called *follicles*. What is a follicle? *Follicles are saclike structures within which the female gametes begin their development.*

How are primary and vesicular follicles anatomically different? *The primary follicle has one or a few layers of cuboidal follicle cells surrounding a larger central developing ovum, whereas the vesicular follicle is surrounded by a capsule of several layers of follicle cells called the corona radiata.*

What is a corpus luteum? *It is a solid glandular structure or a structure containing a scalloped lumen that develops from the ovulated follicle.*

21. What hormone is produced by the vesicular follicle? *Estrogens*

By the corpus luteum? *Estrogens and progesterone*

22. Use the key to identify the cell type you would expect to find in the following structures.

Answer		Structure	Key:
oogonium	1.	forming part of the primary follicle in the ovary	oogonium
primary oocyte	2.	in the uterine tube before fertilization	primary oocyte
secondary oocyte	3.	in the mature vesicular follicle of the ovary	secondary oocyte
ovum	4.	in the uterine tube shortly after sperm penetration	ovum

Multimedia Resources

Format Options

VHS

CD-ROM

DVD

Slides

MATERIALS FOR GENERAL APPLICATION THROUGHOUT THE COURSE

Video

Body Atlas Series (NIMCO, 11-part series, 30 minutes each, VHS, DVD)

- *In the Womb*
- *Breath of Life*
- *Glands and Hormones*
- *Muscle and Bone*
- *Taste and Smell*
- *Sex*
- *The Food Machine*
- *The Skin*
- *Defend and Repair*
- *Visual Reality*
- *Now Hear This*
- *The Brain*
- *The Human Pump*

Human Biology (FHS, 58 minutes, VHS, DVD)

The Human Body: Systems at Work (FHS, 6-part series, 25 minutes each, VHS, DVD)

- *Circulatory System: The Plasma Pipeline*
- *Digestive System: Your Personal Power Plant*
- *Skeletal System: The Infrastructure*
- *Brain and Nervous System: Your Information Superhighway*
- *Muscular System: The Inner Athlete*
- *Respiratory System: Intake and Exhaust*

The Human Body: The Ultimate Machine (CBS, 27 minutes)

The Incredible Human Machine (CBS, 60 minutes, VHS)

The Living Body (FHS, 26-part series, 26-28 minutes each, VHS, DVD)

- *Introduction to the Body: Landscapes and Interiors*
- *The Senses: Skin Deep*
- *Eyes and Ears*
- *Sleep: Dream Voyage*
- *The Urinary Tract: Water!*
- *Digestion: Eating to Live*
- *Breakdown*
- *The Nervous System: Nerves at Work*
- *Decision*
- *Our Talented Brain*
- *Cell Duplication: Growth and Change*
- *Muscles and Joints: Muscle Power*
- *Moving Parts*
- *The Circulatory System: Two Hearts That Beat as One*
- *Breath of Life*
- *Life Under Pressure*
- *Hot and Cold*
- *Hormones: Messengers*
- *Mechanisms of Defense: Accident*
- *Internal Defenses*
- *Reproduction: Shares in the Future*
- *Coming Together*
- *A New Life*
- *Into the World*
- *Aging*
- *Review of Biology: Design for Living*

Mystery of the Senses (CBS, 5-part series, 30 minutes each, VHS)

- *Vision*
- *Smell*
- *Taste*
- *Hearing*
- *Touch*

The New Living Body (FHS, 10-part series, 20 minutes each, VHS, DVD)

- *Bones and Joints*
- *Muscles*
- *Skin*
- *Breathing*
- *Digestion*
- *Blood*
- *The Brain*
- *The Senses*
- *Homeostasis*
- *Reproduction: Designer Babies*

The World of Living Organisms Part 2 (FHS, 8-part series, 15 minutes each, VHS)

- *Genetic Transmission*
- *Genetic Translation*
- *Bones and Muscles*
- *Digestion*
- *Respiration*
- *Circulation*
- *The Kidney*
- *Reproduction*

Software

A.D.A.M.® Interactive Anatomy® 3.0 (AIA, CD-ROM)

A.D.A.M.® MediaPro (AIA, CD-ROM)

A.D.A.M.® Practice (AIA, CD-ROM)

Body Works (WNS, CD-ROM)

The Dynamic Human (CS, CD-ROM)

InterActive Physiology® 8-System Suite (BC, CD-ROM or www.interactivephysiology.com).

Netter's Interactive Atlas of Clinical Anatomy (LP, CD-ROM)

The Ultimate Human Body (ED, CD-ROM)

Visible Human Male (VP, CD-ROM)

MATERIALS SPECIFICALLY RELATED TO PARTICULAR SYSTEMS

Overview of Human Anatomy

Video

Homeostasis (FHS, 20 minutes, VHS)

Homeostasis: The Body in Balance (IM, 26 minutes, VHS)

Homeostasis: The Body in Balance (HRM, 26 minutes, VHS)

The Human Body: The Ultimate Machine (CBS, 27 minutes)

The Incredible Human Machine (CBS, 60 minutes, VHS)

Organ Systems Working Together (WNS, 14 minutes, VHS)

Systems Working Together (WNS, 15 minutes, VHS)

Software

Interactive Atlas of Human Anatomy (ICON, CD-ROM)

The Cell

Video

A Journey Through the Cell (FHS, 2-part series, 25 minutes each, VHS, DVD)

Cell Functions: A Closer Look

Cells: An Introduction

An Introduction to the Living Cell (CBS, 30 minutes, VHS)

Inside the Living Cell (WNS, set of 5, VHS)

Mitosis and Meiosis (UL, 23 minutes, VHS)

The Outer Envelope (WNS, 15 minutes, VHS)

Software

The Cell: Structure, Function, and Processes (HRM, CD-ROM)

Inside the Cell (CE, CD-ROM)

Mitosis (CE, CD-ROM)

Histology

Slides

Basic Human Histology (CBS, microslide sets of eight related 35-mm slides)

Histology Slides for Life Science (BC, 35-mm slides)

Video

Histology Videotape Series (UL, 26-part series, 30 minutes each, VHS)

Software

Eroschenko's Interactive Histology (UL, CD-ROM)

PhysioEX™: Exercise 6B (BC, CD-ROM)

Wards' Histology Collection (WNS, CD-ROM)

Integumentary System

Video

How the Body Works: Skin, Bones, and Muscles (AIMS, NIMCO, 19 minutes, VHS, DVD)

The Senses: Skin Deep (FHS, 26 minutes, VHS, DVD)

Skin (FHS, 20 minutes, VHS)

The Skin (NIMCO, 30 minutes, VHS)

Software

How the Body Works: Skin, Bones, and Muscles (AIMS, NIMCO, CD-ROM, DVD)

Skeletal System

Video

Anatomy of a Runner (Structure and Function of the Lower Limb) (UL, 38 minutes, VHS)

Anatomy of the Hand (FHS, 14 minutes, VHS, DVD)

Anatomy of the Shoulder (FHS, 17 minutes, VHS, DVD)

Bones and Joints (FHS, 20 minutes, VHS, DVD)

Gluteal Region and Hip Joint (UL, 18 minutes, VHS)

How the Body Works: Skin, Bones, and Muscles (AIMS, NIMCO, 19 minutes, VHS, DVD)

The Human Skeletal System (IM, 23 minutes, VHS)

Knee Joint (UL, 16 minutes, VHS)

Movement at Joints of the Body (FHS, 40 minutes, VHS, DVD)

Moving Parts (FHS, 27 minutes, VHS, DVD)

Muscle and Bone (NIMCO, 30 minutes, VHS)

Muscles and Joints: Muscle Power (FHS, 26 minutes, VHS, DVD)

Our Flexible Frame (WNS, 20 minutes, VHS)

Skeletal System: The Infrastructure (FHS, 25 minutes, VHS, DVD)

Skeleton: An Introduction (UL, 46 minutes, VHS)

The Skeletal System (WNS, 15 minutes, VHS)

The Skeleton: Types of Articulations (UL, 16 minutes, VHS)

The Skull Anatomy Series (UL, 9-part series, VHS)

The Thoracic Skeleton (UL, 18 minutes, VHS)

Software

How the Body Works: Skin, Bones, and Muscles (AIMS, NIMCO, CD-ROM, DVD)

Interactive Foot and Ankle (LP, CD-ROM)

Interactive Shoulder (LP, CD-ROM)

Interactive Skeleton: Sports and Kinetic (LP, CD-ROM)

Muscle System

Video

Anatomy of a Runner (Structure and Function of the Lower Limb) (UL, 38 minutes, VHS)

Abdomen and Pelvis (UL, 16 minutes, VHS)

Human Musculature Videotape (BC, 23 minutes, VHS)

Lower Extremity (UL, WNS, 28 minutes, VHS)

Major Skeletal Muscles and Their Actions (UL, 19 minutes, VHS)

Muscles (FHS, 20 minutes, VHS, DVD)

Muscles and Joints: Muscle Power (FHS, 26 minutes, VHS, DVD)

The Skeletal and Muscular Systems (UL, 24 minutes, VHS)

Software

InterActive Physiology® 8-System Suite–Muscular System Module (BC, CD-ROM or www.interactivephysiology.com)

- *Anatomy Review: Skeletal Muscle Tissue*
- *The Neuromuscular Junction*
- *Sliding Filament Theory*
- *Muscle Metabolism*
- *Contraction of Motor Units*
- *Contraction of Whole Muscle*

Nervous System and Special Senses

Video

Anatomy of the Human Brain (FHS, 35 minutes, VHS, DVD)

Animated Neuroscience and the Action of Nicotine, Cocaine, and Marijuana in the Brain (FHS, 25 minutes, VHS, DVD)

Balance (NIMCO, 28 minutes, VHS)

The Brain (FHS, 20 minutes, VHS, DVD)

The Brain (NIMCO, 30 minutes, VHS)

Brain and Nervous System: Your Information Superhighway (FHS, 25 minutes, VHS, DVD)

The Central Nervous System and Brain (IM, 29 minutes, VHS)

Decision (FHS, 28 minutes, VHS, DVD)

The Ear: Hearing and Balance (IM, 29 minutes, VHS)

The Eye: Structure, Function, and Control of Movement (FHS, 54 minutes, VHS, DVD)

The Eye: Vision and Perception (UL, 29 minutes, VHS)

Eyes and Ears (FHS, 28 minutes, VHS, DVD)

Hearing (FHS, 19 minutes, VHS, DVD)

The Human Brain in Situ (FHS, 19 minutes, VHS, DVD)

The Human Nervous System: Spinal Cord and Nerves Videotape (BC, 28 minutes, VHS)

The Human Nervous System: Brain and Cranial Nerves Videotape (BC, 28 minutes, VHS)

Mystery of the Senses (CBS, 5-part series, 30 minutes each, VHS)

- *Vision*
- *Smell*
- *Taste*
- *Hearing*
- *Touch*

The Nature of the Nerve Impulse (FHS, 15 minutes, VHS, DVD)

Nerves and Nerve Cells (NIMCO, 28 minutes, VHS)

The Nervous System: Nerves at Work (FHS, 27 minutes, VHS)

Neuroanatomy (UL, 19 minutes, VHS)

Now Hear This (NIMCO, 30 minutes, VHS)

Optics of the Human Eye Series (UL, 4-part series, VHS)

The Peripheral Nervous System (UL, 29 minutes, VHS)

Reflexes and Synaptic Transmission (UL, 29 minutes, VHS)

The Senses (FHS, 20 minutes, VHS, DVD)

The Senses of Smell and Taste (NIMCO, 28 minutes, VHS)

The Senses: Skin Deep (FHS, 26 minutes, VHS, DVD)

Sheep Brain Dissection (WNS, 22 minutes, VHS)

Sheep Eye Dissection (WNS, 15 minutes, VHS)

Taste (FHS, 23 minutes, VHS, DVD)

Taste and Smell (NIMCO, 30 minutes, VHS)

Visual Reality (NIMCO, 30 minutes, VHS)

Software

InterActive Physiology® 8-System Suite–Nervous System I and II (BC, CD-ROM or www.interactivephysiology.com)

Nervous I	**Nervous II**
Orientation	*Orientation*
Anatomy Review	*Anatomy Review*
Ion Channels	*Ion Channels*
The Membrane Potential	*Synaptic Potentials And Cellular Integration*
The Action Potential	*Synaptic Transmission*

Endocrine System

Video

Body Chemistry: Understanding Hormones (FHS, 3-part series, 50 minutes each, VHS, DVD)

Hormonally Yours

Hormone Heaven?

Hormone Hell

The Endocrine System (IM, UL, WNS, 17 minutes, VHS)

Glands and Hormones (NIMCO, 30 minutes, VHS)

Hormones: Messengers (FHS, 27 minutes, VHS, DVD)

The Hypothalamus and Pituitary Glands (UL, 29 minutes, VHS)

The Neuroendocrine System (IM, UL, 29 minutes, VHS)

Selected Actions of Hormones and Other Chemical Messengers Videotape (BC, 15 minutes, VHS)

Software

InterActive Physiology® 8-System Suite–Endocrine System (BC, CD-ROM or www.interactivephysiology.com)

Orientation

Endocrine System Review

Biochemistry, Secretion, and Transport of Hormones

The Actions of Hormones on Target Cells

The Hypothalamic-Pituitary Axis

Response to Stress

Cardiovascular System

Video

Bleeding and Coagulation (FHS, 31 minutes, VHS, DVD)

Blood (UL, 22 minutes, VHS)

Blood (FHS, 20 minutes, VHS, DVD)

Blood is Life (FHS, 25 minutes, VHS, DVD)

Circulation: A River of Life (WNS, 30 minutes, VHS)

The Circulatory System (IM, 23 minutes, VHS, DVD)

Circulatory System: The Plasma Pipeline (FHS, 25 minutes, VHS, DVD)

The Circulatory System: Two Hearts that Beat as One (FHS, 28 minutes, VHS, DVD)

Human Biology (FHS, 58 minutes, VHS, DVD)

Human Cardiovascular System: Blood Vessels Videotape (BC, 25 minutes, VHS)

Human Cardiovascular System: The Heart Videotape (BC, 25 minutes, VHS)

Life Under Pressure (FHS, 26 minutes, VHS, DVD)

The Mammalian Heart (AIMS, 15 minutes, VHS, DVD)

Pumping Life–The Heart and Circulatory System Video (WNS, 20 minutes, VHS)

The Physiology of Exercise (FHS, 15 minutes, VHS, DVD)

Sheep Heart Dissection Video (WNS, 14 minutes, VHS)

Software

Blood and Immunity (CE, LP, CD-ROM)

InterActive Physiology® 8-System Suite–Cardiovascular System (BC, CD-ROM or www.interactivephysiology.com)

- *Anatomy Review: The Heart*
- *Intrinsic Conduction System*
- *Cardiac Action Potential*
- *Cardiac Cycle*
- *Cardiac Output*
- *Anatomy Review: Blood Vessel Structure and Function*
- *Measuring Blood Pressure*
- *Factors That Affect Blood Pressure*
- *Blood Pressure Regulation*
- *Autoregulation and Capillary Dynamic*

Respiratory System

Video

Breath of Life (FHS, 26 minutes, VHS, DVD)

Breathing (FHS, 20 minutes, VHS, DVD)

The Dissection of the Thorax Series (UL, VHS)

- *Part I. The Thoracic Wall* (23 minutes)
- *Part II. Pleurae and Lungs* (24 minutes)

Human Respiratory System Videotape (BC, 25 minutes, VHS)

Lungs (Revised) (AIMS, 10 minutes, VHS)

The Physiology of Exercise (FHS, 15 minutes, VHS, DVD)

Respiration (FHS, 15 minutes, VHS)

The Respiratory System (UL, 26 minutes, VHS)

Respiratory System: Intake and Exhaust (FHS, 25 minutes, VHS, DVD)

Thorax (UL, 22 minutes, VHS)

Software

InterActive Physiology® 8-System Suite–Respiratory System
(BC, CD-ROM or www.interactivephysiology.com)

Anatomy Review: Respiratory Structures

Pulmonary Ventilation

Gas Exchange

Gas Transport

Control of Respiration

Digestive System

Video

Breakdown (FHS, 28 minutes, VHS, DVD)

Digestion (FHS, 20 minutes, VHS, DVD)

Digestion: Eating to Live (FHS, 27 minutes, VHS, DVD)

Digestive System (WNS, 14 minutes, VHS)

Digestive System: Your Personal Power Plant (FHS, 20 minutes, VHS, DVD)

The Food Machine (NIMCO, 30 minutes, VHS)

The Guides to Dissection Series (UL, VHS)

Group V. The Abdomen (6 parts, 88.5 minutes total)

The Human Digestive System (AIMS, 18 minutes, VHS)

Human Digestive System Videotape (BC, 33 minutes, VHS)

Passage of Food Through the Digestive Tract (WNS, 8 minutes, VHS)

Urinary System

Video

Human Urinary System Videotape (BC, 23 minutes, VHS)

Kidney Functions (AIMS, 5 minutes, VHS, DVD)

The Kidney (FHS, 15 minutes, VHS)

The Urinary Tract: Water! (FHS, 28 minutes, VHS, DVD)

Software

InterActive Physiology® 8-System Suite–Urinary System (BC, CD-ROM or www.interactivephysiology.com)

Anatomy Review

Glomerular Filtration

Early Filtrate Processing

Late Filtrate Processing

InterActive Physiology® 8-System Suite–Fluids and Electrolytes (BC, CD-ROM or www.interactivephysiology.com)

Introduction to Body Fluids

Water Homeostasis

Electrolyte Homeostasis

Acid/Base Homeostasis

Reproductive System

Video

The Guides to Dissection Series (UL, VHS)

Group VI. The Pelvis and Perineum (4 parts, 64 minutes total)

Human Biology (FHS, 58 minutes, VHS, DVD)

The Human Female Reproductive System (UL, 29 minutes, VHS)

The Human Male Reproductive System (UL, 29 minutes, VHS)

Human Reproductive Biology (FHS, 35 minutes, VHS, DVD)

Human Reproductive System Videotape (BC, 32 minutes, VHS)

Reproduction: Shares in the Future (FHS, 26 minutes, VHS, DVD)

Multimedia Resource Distributors

appendix **B**

Note: *Distribution company names and web addresses change frequently.*

AIMS AIMS Media
9710 Desoto Avenue
Chatsworth, CA 91311-4409
800-367-2467/818-773-4300
www.aimsmultimedia.com

BC Benjamin Cummings
1301 Sansome Street
San Francisco, CA 94111-2525
800-950-2665/415-402-2500
www.aw-bc.com

CBS Carolina Biological Supply Company
2700 York Road
Burlington, NC 27215
800-334-5551
www.carolina.com

CE CyberEd, Inc.
P.O. Box 3480
Chico, CA 95927-3480
888-318-0700/530-899-1212
www.cybered.net

FHS Films for the Humanities and Sciences
P.O. Box 2053
Princeton, NJ 08453
800-257-5126/609-419-8000
www.films.com

HRM HRM Video
41 Kensico Drive
Mount Kisco, NY 10549
800-431-2050
www.hrmvideo.com

ICON Icon Learning Systems
295 North Street
Teterboro, NJ 07608
800-631-1181
www.netterart.com

IM Insight Media
2162 Broadway
New York, NY 10024
800-233-9100/212-721-6316
www.insight-media.com

LP Laser Professor
351 Lakeside Lane, Suite 308
Houston, TX 77058
800-550-0335/281-333-5550
www.laserprofessor.com

NIMCO NIMCO, Inc.
P.O. Box 9
102 Highway 81 N
Calhoun, KY 42327-0009
800-962-6662
www.nimcoinc.com

UL United Learning
1560 Sherman Avenue, Suite 100
Evanston, IL 60201
800-323-9084
www.unitedlearning.com

WNS Ward's Natural Science
5100 West Henrietta Road
Rochester, NY 14692-9012
800-962-2660/585-359-2502
www.wardsci.com

ADDITIONAL MULTIMEDIA RESOURCES

A.D.A.M. Software, Inc.
1800 RiverEdge Parkway, Suite 100
Atlanta, GA 30328
770-980-0888
http://education.adam.com

Ambrose Video
145 West 45th Street, Suite 1115
New York, NY 10036
800-526-4663/212-768-7373
www.ambrosevideo.com

Anatomical Chart Company
4711 Golf Road
Suite 650
Skokie, IL 60076
847-679-4700/847-679-9155
www.anatomical.com

Annenberg/CPB Project
The Corporation for Public Broadcasting
P.O. Box 2345
S. Burlington, VT 05407-2345
1-800-LEARNER
www.learner.org

Communication Skills, Inc.
49 Richmondville Avenue
Westport, CT 06880
800-824-2398/203-226-8820
www.communicationskills.com

Concept Media
PO Box 19542
Irvine, CA 92623-9542
800-233-7078
www.conceptmedia.com

Denoyer Geppert Company
PO Box 1727
Skokie, IL 60076-8727
800-621-1014/866-531-1221
www.denoyer.com

Educational Images, Ltd.
PO Box 3456 Westside Station
Elmira, NY 14905-0456
800-527-4264/607-732-1183
www.educationalimages.com

Edumatch, Inc.
86 West Street
Waltham, MA 02451-1110
800-637-0047
www.edumatch.com

Eli Lilly & Company, Medical Division
Lilly Corporate Center
Indianapolis, IN 46285
317-276-2000
www.lilly.com

EME Corporation
581 Central Parkway
P.O. Box 1949
Stuart, FL 34995
800-848-2050/772-219-2206
www.emescience.com

Encyclopedia Britannica Educational Corporation
310 South Michigan Avenue
Chicago, IL 60604
800-323-1229/312-294-2104
www.britannica.com

Fisher Scientific Education
3970 John's Creek Court, Suite 500
Suwanee, GA 30024
800-766-7000/770-871-4726
www.fishersci.com

Hawkhill Associates, Inc.
125 East Gilman Street
Madison, WI 53703
800-422-4295/608-251-3924
www.hawkhill.net

HW Wilson Company
(publishes *Biological and Agricultural Index Plus, General Science Database*)
950 University Avenue
Bronx, NY 10452
800-367-6770/718-588-8400
www.hwwilson.com

Milner-Fenwick, Inc.
2125 Greenspring Drive
Timonium, MD 21093
800-432-8433
www.milner-fenwick.com

Nebraska Scientific
3823 Leavenworth Street
Omaha, NE 68105
800-228-7117/402-346-7214
www.nebraskascientific.com

National Geographic Film Library
1145 17th NW
Washington, D.C. 20036
877-730-2022/202-429-5755
www.natgeostock.com

National Geographic Society
P.O. Box 10041
Des Moines, Iowa 50340-0041
888 CALL NGS (888 225 5647)
www.nationalgeographic.com

Public Broadcasting Service
1320 Braddock Place
Alexandria, VA 22314
www.pbs.org

Phoenix Learning Group
2349 Chaffee Drive
St. Louis, MO 63146
800-221-1274
www.phoenixlearninggroup.com

Pyramid Media
P.O. Box 1048/WEB
Santa Monica, CA 90406
877-730-2022/202-429-5755
www.pyramidmedia.com

Queue, Inc.
1450 Barnum Avenue, Suite 207
Bridgeport, CT 06610
800-232-2224/800-775-2729
www.queueinc.com

RAmEx Ars Medica, Inc.
1511 Sawtelle Blvd. #232
Los Angeles, CA 90025-3206
800-633 9281/310-826 9674
www.ramex.com

Riverdeep Interactive Learning
500 Redwood Boulevard
Novato, CA 94947
888-242-6747/415-763-4700
www.riverdeep.net

Research Systems, Inc.
4990 Pearl East Circle
Boulder, CO 80301
303-786-9900
www.rsinc.com

Scientific American
415 Madison Avenue
New York, NY 10017
800-333-1199/212-754-0550
www.sciam.com

Thomson/Gale
(publishes *Video Source Book*)
27500 Drake Road
Farmington Hills, MI 48331
800-877-4253/248-699-4253
www.galegroup.com

University of California Extension
Center for Media and Independent Learning
2000 Center Street, Fourth Floor
Berkeley, CA 94704-1223
510-642-0460/510-643-9271
http://ucmedia.berkeley.edu

Videodiscovery, Inc.
920 N. 34th
Seattle, WA 98103
800-548-3472/206-285-5400
www.videodiscovery.com

Visible Productions
201 Linden Street, Suite 301
Fort Collins, CO 80524
800-685-4668
http://visiblep.com

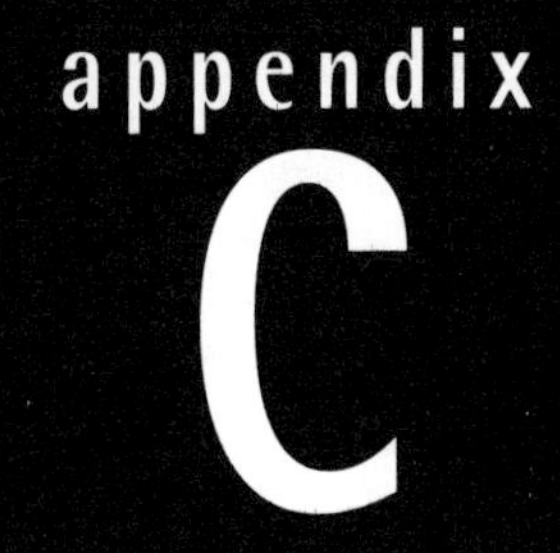

Supply Houses

This is a partial list of suppliers of equipment, animals and chemicals, and should not be considered a recommendation for these companies. Many supply companies have regional addresses. Only one address is listed below.

Aldrich Chemical Company, Inc. *
P.O. Box 2060
Milwaukie, WI 53201
800-558-9160
www.sigma-aldrich.com

Carolina Biological Supply Company
2700 York Road
Burlington, NC 27215
800-334-5551
www.carolina.com

Fisher Scientific
3970 John's Creek Court, Suite 500
Suwanee, GA 30024
800-766-7000/770-871-4726
www.fishersci.com

ICN Biochemicals
1263 South Chillicothe Road
Aurora, OH 44202-8064
800-854-0530
www.icnbiomed.com

Intelitool (Phipps & Bird)
P.O. Box 7475
Richmond, VA 23221-0475
800-955-7621
www.intelitool.com

LabChem, Inc.
200 William Pitt Way
Pittsburgh, PA 15238
412-826-5230
www.labchem.net

Nasco
901 Janesville Avenue
P.O. Box 901
Fort Atkinson, WI 53538-0901
800-558-9595
www.enasco.com

Sigma Chemical Company *
P.O. Box 14508
St. Louis, MO 63178-9974
800-325-3010
www.sigma-aldrich.com

VWR International, Inc.
200 Center Square Road
Bridgeport, NJ 08014
800-932-5000
www.vwrsp.com

Ward's Natural Science
5100 West Henrietta Road
Rochester, NY 14692-9012
800-962-2660/585-359-2502
www.wardsci.com

* The Aldrich and Sigma brands are both sold through Sigma-Aldrich, Inc.